THE SPEED OF LIGHT

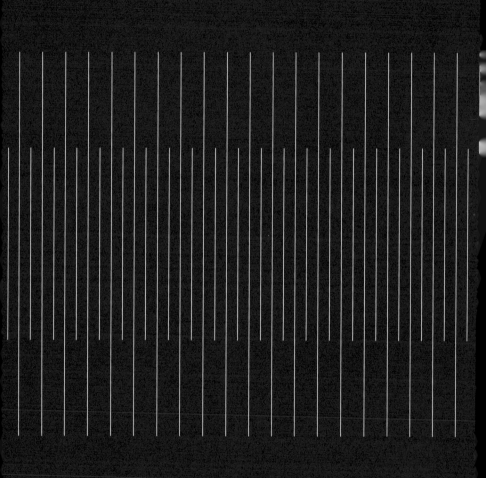

INDIANA UNIVERSITY PRESS   *Bloomington & Indianapolis*

THE SPEED OF LIGHT

CONSTANCY + COSMOS

DAVID A. GRANDY

This book is a publication of

Indiana University Press
601 North Morton Street
Bloomington, IN 47404-3797 USA

http://iupress.indiana.edu

Telephone orders   800-842-6796
Fax orders   812-855-7931
Orders by e-mail   iuporder@indiana.edu

∞The paper used in this publication meets the minimum requirements
of American National Standard for Information Sciences—Permanence
of Paper for Printed Library Materials, ANSI Z39.48-1992.

Manufactured in the United States of America

Library of Congress Cataloging-in-Publication Data

Grandy, David.
   The speed of light : constancy and cosmos / David A. Grandy.
      p. cm.
   Includes bibliographical references and index.
   ISBN 978-0-253-35322-1 (cloth : alk. paper) — ISBN 978-0-253-22086-8 (pbk. :
alk. paper)
   1. Light—Speed. 2. Space and time. 3. Time—Philosophy. I. Title.
   QC407.G73 2009
   535—dc22
                              2008043194

1 2 3 4 5 14 13 12 11 10 09

To my parents, Grant and Barbara Grandy

For as the eyes of bats are to the light of day, so is the intellect of our soul to the objects which in their nature are most evident of all.

ARISTOTLE

CONTENTS

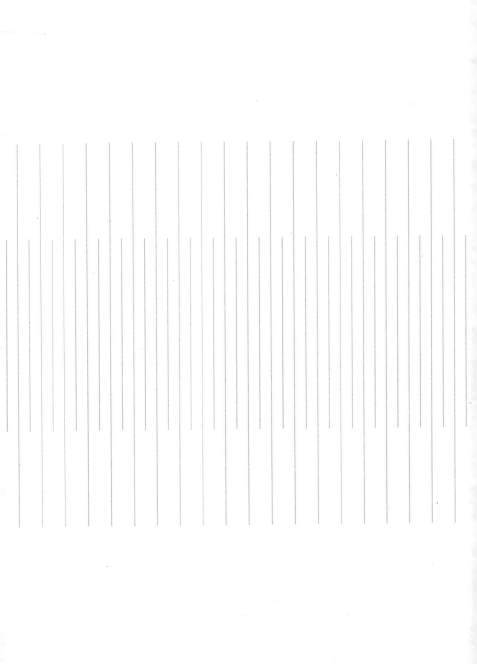

ACKNOWLEDGMENTS

Every author grows blind to the manuscript he or she writes, and so fresh eyes are always welcome. I thank an anonymous reviewer at Indiana University Press for meticulously correcting errors and proposing new directions and emphases. I similarly thank Sienna Dittmer, Ace Sorensen, Chad McKell, and Marc-Charles Ingerson for reading the text, identifying weaknesses, producing illustrations, and patiently listening to my ideas. I am also appreciative of the faculty seminars I've participated in at Brigham Young University. Directed and taught by Travis Anderson, James Faulconer, and Mark Wrathall, these weekly gatherings sparked my interest in phenomenology and allowed me to think more broadly about puzzling issues in science. I also wish to thank, or perhaps apologize to, students in a recent undergraduate philosophy course who were assigned to work their way through Alfred North Whitehead's *Science and the Modern World*. I say "apologize to" because Whitehead is not easy to read, but the endeavor stretched me in new ways, as I hope it did my students as well. In this regard let me thank David Paulsen, whose invitation to write a scholarly article on process thought led me deeper into Whitehead's contribution to modern philosophy.

Among the editorial staff at Indiana University Press I express thanks to Robert Sloan, June Silay, Carol Kennedy, Chandra Mevis, and Miki Bird. Always kind and professional, they made it easy for me to keep thinking about the manuscript even though I was involved in other projects. I also thank Dan Burton for his willingness to trade ideas with me. I remember fondly our graduate school discussions, many of which occurred under the auspices of the short-lived "Space and Time Club" whose sole characteristic was a friendship-based quest to grasp modern physics and its philosophical ramifications. I think of this book as an outgrowth of those discussions.

Finally, I thank my family, particularly my wife Janet, for supporting me in my endeavors. Without this backdrop of love and support I would not have finished, or even started, this book.

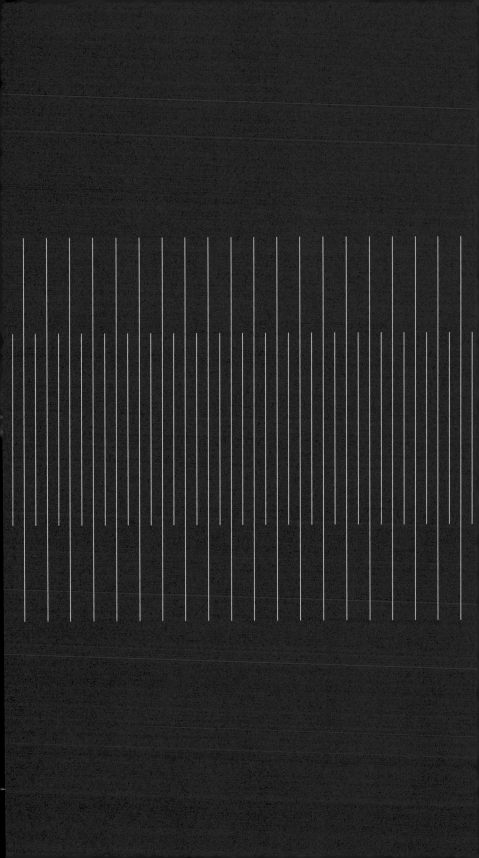

THE SPEED OF LIGHT

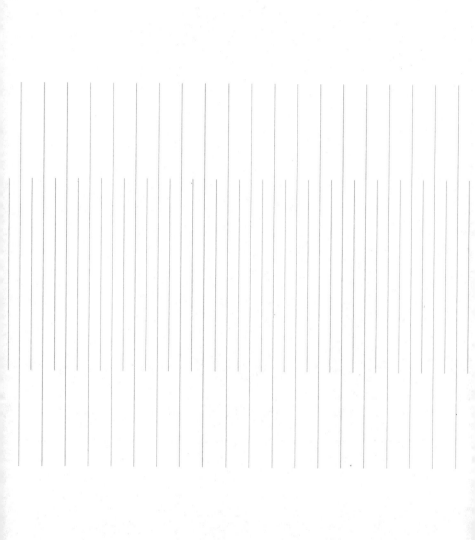

INTRODUCTION

Years ago I learned something that has bothered me ever since: the speed of light is constant in all inertial reference frames.[1] This is a simple fact of science, easy to state and easy to remember. Its scientific implications, however, are enormous. What is more, from a commonsensical point of view, it makes no sense. Here is how David Mermin takes up the question: "How can this be? How can there be a speed c with the property that if something moves with speed c, then it must have the speed c in any inertial frame of reference? This fact—known as the *constancy of the speed of light*—is highly counterintuitive. Indeed, 'counterintuitive' is too weak a word. It seems downright impossible."[2]

Daily experience is all about motion, but only light moves absolutely. That is, when it comes to measuring and calculating speed, nothing but light returns the same value to all observers. Normally such calculation involves adding and subtracting velocities. If you and I bump into each other, each while walking two miles an hour relative to the earth's surface, the effective impact speed is four miles an hour. If you and I are walking in the same direction, you at five miles an hour and I at three, the relative difference between our speeds is two miles an hour.

This is all very straightforward, and it is the way we think about speed, even if we normally don't carry out the calculations and even if, most of the time, we bring all speeds onto the same page, so to speak, by using the earth as a common benchmark. Things move relative to each other, and tallying the speed of one thing involves knowing the speed of the other. What would it mean, then, to encounter something that is perfectly indifferent to the speed of other things, something whose own speed cannot be diminished or augmented by any other speed?

Light is that something. It is its own speed, not a speed that partly depends on the motion of something else. This, of course, makes light strange and unique, but to fully appreciate its nature, we must go further than that. I believe that light is more than strange: it is constitutive of reality, of our own being, and therefore not really strange but deeply familiar. It is quietly resident in the universal shape of things, not some oddball phenomenon off on the fringe. So if light does not play the everyday game of relative speeds, it is because it structures that game in the first place. Its absolute speed betokens a deep cosmic fixedness—a constancy—within which relative change can occur, just as fixed platforms—stages and playing fields—enable the dynamic excitement of dramas and sporting events.

Hans Reichenbach had something like this in mind in explaining the significance of light in modern physics:

> Clocks and yardsticks, the material instruments for measuring space and time, have only a subordinate function. They adjust themselves to the geometry of light and obey all the laws which light furnishes for the comparison of magnitudes. One is reminded of a magnetic needle adjusting itself to the field of magnetic forces, but not choosing its direction independently. Clocks and yardsticks, too, have no independent magnitude; rather, they adjust themselves to the metric field of space, the structure of which manifests itself most clearly in the rays of light.[3]

One may think of light as a manifestation of cosmic unity, a primal integrity that permits the emergence of plurality and difference within a *single* context or system that never shatters under the weight of multiplicity. This resistance to absolute difference or otherness, this something that keeps various things quietly and collectively *intact* while allowing them to show up as apparently independent entities, is what we have come to call light. If this is right, and I believe it is, we can then begin to understand why the speed of light is constant in all reference frames.

What is at issue here, however, is something larger than the puzzling fact of light speed constancy. That fact is just one of many ways that modern physics breaks the frame of everyday reality, and I use it as a pathway into the proposition that the universe, struck from light at the big bang singularity, conserves the spaceless, timeless character of that first moment. Hence, first light, the notion that light conserves the moment of creation.

Here *first* means originary or primary, but not in the sense of being first in time. To be first in time implies that time is already underway, that time itself is first, but the light of modern physics is ontologically prior to time in the sense that it originates spacetime creation but does not fully participate in it. To adapt a line from Jacques Derrida, light is a "revealability more originary than revelation."[4] Or as Hans Blumenberg expressed it, light is "the 'letting-appear' that does not itself appear."[5] It sets the cosmic revelation—the cosmos—in spacetime motion while remaining, in some ways, indifferent to that motion.[6]

Light speed constancy is a manifestation of that indifference: this is the springboard to the larger thesis that light is the timeless miracle of creation that informs and structures every temporal, post-creation event. Stated this way, however, the thesis sounds extravagant, and that is why we must begin with something factual and concrete, the puzzle of light speed constancy, and see where it leads us.

Note that I am not worried about the speed of light per se. My concern is with light speed *constancy*. So if the measured value of the speed of light were slightly improved, that would have no bearing on my argument. Even if that were to happen, all observers, using the same measurement technology, would still register the same speed value for a light ray. And it would not matter how differently they might be moving relative to the ray. The speed of light would be constant across all perspectives. Put another way, if measurement differences *were* to show up, they would be attributed to measurement error, not to the differing motions of the observers. This has been the consensus of science, and one of its best confirmed postulates, since the early decades of the twentieth century.

Over the years I have consulted dozens of texts but failed to find one that comprehensively addresses the question that sets the stage for this book: Why is the speed of light constant for all observers?

Physics texts typically state the fact of light speed constancy without trying to explain it; more pressing, evidently, is the need to get on with calculations and applications. My sense is that most scientists are content to let it serve an axiomatic role, and this is how Albert Einstein introduced it in his first paper on special relativity—not as a verified fact but as one of two axioms from which inferences could be drawn. But axioms, as traditionally employed, are self-evident posits, and light speed constancy is anything but self-evident. Today, of course, it is both an axiom and an empirical fact, but its experimental verification makes it no less puzzling.

We can, of course, write it off as unfathomable, but why assume the intellectually modest approach before trying something more adventurous. If light speed constancy is a fact about the world, and if it opens a window on the world's deep structure, then perhaps it is worth trying to look through that window. I propose that this constancy denotes our inevitable involvement in nature. It marks the mutual interpenetration of mind and matter; it is a place—an interface—where these two entities dissolve into one another. We cannot, consequently, overtake it—disrupt that constancy—as we overtake things less familiar or intimate. If we were interlopers in the cosmos, perhaps we could overtake every aspect of nature in a clear and objective way. Since we aren't, however, certain complicities or intertwinings work against our objective apprehension of the world. Light shows up in such places, at the horizon or interface of physical and perceptual reality, and this is why it is anomalous. More to the point, light speed constancy is a consequence of the fact that upon measuring the speed of light, we are already complicit with light. We are already incorporated into a cosmos whose definitional structure, or some aspect thereof, cashes out as a single and singular value—an absolutely constant value—known as the speed of light.

Modern physics has debunked the notion that we can observe nature without participating in it. But even here it seems reflexive among physicists to assume we are outside looking in. The true significance of relativity theory and quantum theory, I believe, is that there is no outside. We are in the system, and not just as loose parts rattling about. Instead we are integrated or patterned into nature and therefore share in its limitations. At the same time, however, the universe is open precisely because that patterning goes all the way down, so to speak, and therefore knows no boundaries or limits. My perceptual faculties, particularly sight, reach outward by virtue of a patterning so elemental as to distribute me throughout the whole. And subtly textured throughout, I have a wide perceptual presence. Not only that, but the universe does not loom up as a cage or some sort of alien confinement. It may be imagined as such, but perceptually speaking, it is unbounded. Looking skyward, I cannot see to the end of things, and this is because there is no point at which the world's physiology, so to speak, departs from my own.

In making these claims I am resorting to phenomenology. The modern turn, or return, to perceptual experience as a way of grasping the world goes back to Johann Wolfgang von Goethe in the early nineteenth century. Nearly a hundred years would pass, however, before Edmund Husserl took up the project more rigorously, and others have since contributed to the phenomenological movement. I propose that light speed constancy invites and even cries out for phenomenological analysis. Like certain other puzzling issues in physics, that constancy segues naturally into a consideration of the observer's role in the shaping of physical reality. But, as suggested earlier, physicists typically shy away from philosophically sticky issues, even though most realize that their discipline is in philosophical turmoil. I suggest that some of the turmoil can be tamed, or at least made sense of, by reading light speed constancy back into everyday perceptual experience, and phenomenology can help out in this endeavor. That is one way to get started on the question of why the speed of light is constant in all inertial reference frames.

Making the same point in a different way, let me note that as one does physics it is easy and all but compulsory to begin thinking of light speed constancy as a constant of nature having little or no relevance to everyday experience. The initial shock of light speed constancy wears off as c—shorthand for the speed of light in a vacuum—shows up in fundamental equations, which then command one's thinking about the world. This is good in a way, for it establishes a point central to my argument—the speed of light is an integral aspect of nature. What goes unasked, however, is whether it is so integral as to figure into mundane visual experience, which, after all, is light mediated. I believe it does.

The early chapters take up the question of light speed constancy from a scientific point of view. I explain why Einstein proposed the idea and how it reshapes our understanding of space and time. The important point here, and one that is rarely emphasized, is that with Einstein light speed constancy became a *measured* constancy, measurement being an undeniable aspect of our involvement in nature. Put another way, at least with respect to its speed, the circuit of light completes itself in perceptual experience, not, as was previously believed, independently of human observation. This link with the observer carries through into the theory itself, where certain properties of objects are not absolute or observer-independent. Rather an object's length, say, is constituted by the compound system of observer and object.

I argue that light speed constancy begins to make sense if one is willing to drop the assumption that light is something exclusively "out there," something apart from our experience of light. As the book progresses, I try to drive this point home in different ways. Thinkers since Plato have surmised that light drops out of sight to give us sight; it is an invisible clarity that affords clear vision of other things. It thus seems to be more a principle of seeing than something to be seen. If this is correct, it follows that light is a matter of experience, and it stays in play as experience even as we do our best to reduce it to rule and explanation. In our abstractive distancing of light, we never quite cut the thread that leads back to light's experiential immediacy.

Indeed, given light's primordiality—a feature of both religious and scientific cosmology—one might insist that experiential light is older than and therefore originative of theoretical light, the post-experience or post-experiment urge to map, analyze, and explain. After remarking that "[p]hilosophy lives in words, but truth and fact well up into our lives in ways that exceed verbal formulation," William James stated: "There is in the living act of perception always something that glimmers and twinkles and will not be caught, and for which reflection comes too late."[7] I am proposing that even though reflection comes too late, the living act of perception is not thereby completely arrested or put out of play. It lives on in the reflective process, all the while making that process a game of catch-up. Indeed it gives birth to that process. "First he will see the sun and then he will reason about it," wrote Plato in describing the prisoner released from a cave of shadows.[8] Theoretical light, the light we come to reason about after witnessing light, is still, at bottom, experiential, still an instance of light staying in play.

To stay in play is to stay alive, as it were, to the situation at hand. Light, I believe, is that aspect of reality that cannot be put out of play by reductionistic analysis. Why? Because it fosters that sort of thinking in the first place. This may not be a very original stance—it recalls Hopkins's "There lives the dearest freshness deep down things"[9]—but Einsteinian physics deepens it in surprising ways. Because of light speed constancy, the speed of light, conceptually speaking, is not fully graspable. To be sure, we always *measure* light at a constant velocity. But when we try to *imagine* its motion within its own economy or reference frame, the puzzle of light's motion, first encountered as unfailing constancy, is raised to a higher power by the relativistic effects of time dilation and length contraction. What emerges is the thought that in some sense light does not move at all: being indifferent to space and time, it cannot be plotted accordingly. With this fact in mind, some have characterized light as ageless or atemporal and recognized that it makes "zero-interval linkages between events near and far."[10] While this is an abstruse claim, I believe it is the very stuff of visual experience and call upon James Gibson and Maurice Merleau-Ponty while making an argument to that effect.

Seeking to anchor the study of vision to pre-reflective experience, Gibson and Merleau-Ponty discovered that aspect of light that couples and conjoins rather than separates. One routinely imagines light in intermediate space, moving between object and observer. This is only half the story, however, and the latter half at that. The first half, which furnishes the foundation for what comes after, subsists in what Merleau-Ponty called the magic or "delirium" of visual experience.[11] We open our eyes and have immediate visual apprehension of the world. "To see is *to have at a distance*," says Merleau-Ponty, marveling that this having is completely unasked-for and completely at odds with the attitude that we are localized beings cut off from the rest of the world.[12] How can this attitude explain everything, he asks, when we, from the very start, have visual attachment to objects that we then, upon theoretical reflection, determine to be detached from us?

Of course it is true that in some sense distant objects are detached from us, for distance implies detachment. But the primitive experience of seeing, which Merleau-Ponty and Gibson seek to recover, diverges toward expansive conjunction—even coincidence—with those objects. "The first time we see *light*," wrote William James, "we *are* it rather than see it."[13] At a level almost too elemental for ordinary comprehension, light is an opening or window on existence, and we *are* the opening. A person who values a window just for its material properties fails to grasp the idea of a window. Windows begin with a few material constituents having well-marked boundaries—wooden frame, glass pane, and so on—but they end, or, more correctly, fail to end, with light. We also, despite similar material limitations, fail to end with light. Those limitations give us structure and get us started, but, thanks to light, we become more than those limitations, just as a window becomes more—much more—than its material properties.

What I don't want to do is suggest that light is something special or unique. To some extent, perhaps, that can't be avoided. But if readers leave with just that impression, they will have missed the larger point: light is not one thing among many others but rather radiance born of cosmic coherence and relationality. Light per se is never seen; its nature is to flash into existence while announcing other things. It is always a matter of conjunction or coupling, an illuminating click when two or more things click together. Chapters 5 and 6 address this point. Chapter 11 carries the point further by looking at Martin Heidegger's concept of *Lichtung*, or clearing.

Among phenomenologists, Heidegger has commented most perceptively about the light-like clearing or opening—the window, as it were—that constitutes human existence. It would be easy to call this *light*, but that would not capture what Heidegger is talking about. For most people, light is distinct from human existence; furthermore, it has a bottom. It, so this thinking goes, can be explained by science, notwithstanding the puzzles about light that have emerged in modern physics. But this is not Heidegger's outlook. He wants to get back to whatever enables science and philosophy, and for him that is an originary light or clearing (*Lichtung*) that is constitutive of our being in the world.

This then is the explanation for light speed constancy: too close to home or informative of our nature to be objectified, light is not distinct from our experience of light. When we therefore measure the speed of light, we are measuring something that remains in play, something that keeps moving with us on our experiential way. Put differently, to measure the speed of light is to measure something about the way we ourselves are measured or blended into the cosmos, and that universal blending pre-decides our measurement of light speed in favor of a universal or constant value. This is a constant of nature inclusive of our nature.

And to complete the argument: light speed constancy is inclusive of, even indistinguishable from, the first moment of creation, the originary event when spacetime bounds were set to what could and could not occur thereafter. I say "indistinguishable from" because, as noted earlier, time does not figure into light's intrinsic nature. On the scene before space and time got underway, light remains outside their embrace, as evidenced by Einsteinian relativity. So, for light one moment is indistinguishable from another, or, as J. T. Fraser puts it, "All instants in the life of the photon are simultaneous."[14]

The same claim can be made for points in space, but the following is a more intuitive way of grasping how light spacelessly conserves the big bang singularity in everyday experience. That singularity was centrally located; as the originary event, the genesis of space and time, the expansion point from which the cosmos materialized, it could not be otherwise. Similarly, owing to light speed constancy, I am always centrally located, always a kind of singularity around which the world pivots. If you produce a spark at your location, its light moves away from you omni-directionally at light speed, thus situating you right at the center of things. This is not surprising. The surprise comes when I perceive that spark. Although I may be far from you and even speeding away from you, light from the spark envelops me just as it envelops you: it expands away from me omni-directionally at a single speed—the speed of light. I also am right at the center of things, perfectly balanced in all directions, and so it goes for every observer. In sum, first light, the light of creation, is not something tucked away in the past and steadily growing more remote. Rather, in virtue of its ontological priority, light remains eternally operative in the spacetime cosmos. For us, of course, it shows up as everyday light, but it is still first light, the originary light of creation that structures the universe and our experience thereof.

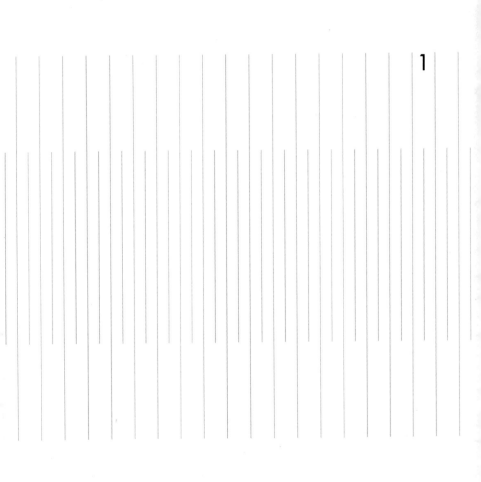

1

SPACE, TIME, AND LIGHT SPEED CONSTANCY

The speed of light is really not like other speeds; it is the
quantity that has the most fundamental implications for the
structure of space and time, or, better yet, of space-time.

Harald Fritzsch, An Equation that Changed the World:
Newton, Einstein, and the Theory of Relativity

Space and time are notoriously hard to define. That, perhaps, is
because they seem mostly blank or featureless. Space appears expansive and three-dimensional, time linear and irreversible. We do not
actually observe these qualities, however. We infer them from events
that supposedly occur *in* space and time. It is as if space contains
events and time keeps them in process. Not only that, but events
unfold in space and time while space and time remain blithely
unaffected by those events.

This is Isaac Newton's version of space and time. He insisted
that "material bodies are indifferent to motion"[1] because real or absolute space and time embody mathematical truths.[2] As such they
offer no resistance to physical bodies, and their scientific value lies
in their metrical properties. They mark times and locations of events

as lightly as a zip code marks a neighborhood—that is, hardly at all. It is almost as if space and time are ideas in God's mind, and in fact Newton, a deeply religious man, so characterized them.[3]

Albert Einstein's outlook is very different. He incorporates space and time into the physical universe. As a result, moving objects are not indifferent to their motion. As they go faster relative to stationary observers, they "warp" or "bend" in ways that keep them from reaching the speed of light. And since these effects embody spacetime changes, one may say that spacetime itself is bent or warped.

Three prominent physicists sum up this coupling of physical matter and space (more precisely, spacetime) in the following way: "Space acts on matter, telling it how to move. In turn, matter reacts back on space, telling it how to curve."[4] Like two hands drawing one another, matter and spacetime mutually call forth each other's properties and modes of action. This interactive coupling begins to make sense once one realizes that physical bodies are not embedded in spacetime as foreign objects: they are material realizations of particular spacetime values (such as length). The spacetime they negotiate also informs them, is part of their reality. Motion consequently is not really a matter of objects passing blithely through an unresponsive or indifferent spacetime field. In John Wheeler's words, "spacetime geometry is not a God-given perfection standing high above the battles of matter and energy, but is itself a participant in the world of physics."[5] That geometry informs physical matter—is in fact inseparable from it—and this means that moving objects are occasions of spacetime self-interference. Stresses arise as spacetime—material embodiments of spacetime—moves through spacetime.

These analogies are far from perfect, but they offer some insight into the dynamic of relativistic motion, and here I cannot resist a slightly grander analogy. The Australian Aborigine creation story describes a flat and featureless topography that was then marked or defined by ancestral spirits during Dream Time, so that hills, valleys, trails, and ponds emerged. Newtonian space is flat and featureless, while Einsteinian space is often warped, curved, and otherwise marked or altered by physical objects. Flat Newtonian space invites the proposition of infinite velocity—what is there about it to slow objects down? But Einsteinian space, with its ripples and irregularities, presents us with built-in speed bumps. Moreover, since material objects are woven into the fabric of Einsteinian spacetime, it seems right that they are not free to travel infinitely fast. Their motion or speed is limited by the fabric itself, by its elastic properties, as it were. Thus, Einstein's idea of a limiting velocity begins to make sense.

But where, exactly, does the idea come from? As near as we can tell, Einstein's sense of beauty or symmetry guided his thinking as he developed his new theory of motion. It is commonplace to attribute the theory to Michelson and Morley's failure to detect the universal ether, but that is just one of many considerations. The story begins with deceptively simple questions about motion: what constitutes motion and how do we know when a body is, in fact, moving?

From Aristotle to Newton

At first blush, motion seems the most obvious of all phenomena. We all know when we are moving and when we are at rest. Or at least we know if we tacitly assume—as all of us do most of the time—that the earth is at rest. The earth then becomes a benchmark against which motion can be ascertained. The hitch here, of course, is that we also know, in an academic way, that the earth is not at rest. Experience or common sense tells us one thing while science tells us another.

Before Copernicus, science coincided with experience on this issue. Aristotle had reasoned that the earth was immobile by virtue of its *gravitas*, or heaviness. In his universe, earthy matter naturally tended centerward, and the earth's sphericity was the result of omnidirectional clumping. Since there was only one cosmic center, the earth was held fast by its own innate properties. Not only that, but the earth, being at rest, became the obvious reference point from which to measure motion. Objects at rest in the cosmos were those at rest with respect to the earth; objects in motion were those moving with respect to the earth.

Copernicus, of course, put the earth in motion around the sun. Later Newton, drawing on the work of Galileo and Descartes, explained why we do not experience the earth's motion even though, by conventional standards, it is very swift. His first law of motion states that a body at rest or in motion remains in that state until acted upon by an outside force. This may sound innocuous, but it is powerful science; it also embodies a new attitude toward nature.

Whereas Aristotle had supposed that material bodies such as rocks and stars are sentient, Newton deemed them lifeless. Thus his first law of motion, also called the principle of inertia, assumed that bodies were constitutionally incapable of self-activated motion. They could not, as Aristotle had reasoned, feel themselves in the cosmos—feel or sense where they were and then move, if necessary, to places cognate with their being. A rock thrown in the air, said Aristotle, eventually returns to its natural place, the earth. Stars move in circular orbits, befitting their eternal (beginningless and endless) nature. By contrast, rocks and stars in Newton's universe had no capacity to "incline toward" (to use the old language) a particular location or kind of motion. Being blank within (lacking sentience or life principle), they were *inert*, a word that means both lifeless and sluggish.

Newton called his law of inertia a "force of inactivity."[6] A body remained in a particular state of rest or motion because it was reactive in a purely mechanical way. It had no capacity to sense its place or motion in the cosmos; indeed it had no capacity to sense itself. Lacking all this, it also lacked the capacity to initiate changes in motion, to be active on its own (non-existent) behalf.

This is one reason that in Newtonian physics material bodies are said to be indifferent to their states of motion. Another is that Newton's universe is qualitatively homogenous and therefore void of places and modes of action uniquely suited to particular elements or kinds of bodies. No place, not even the starry heavens, stands out as innately better than any other. What counts (literally) is numbers or quantitative distinctions. As suggested earlier, Newton regarded space and time as mathematical fixtures that allowed humans to chart events—their positions and times—with scientific precision. Heavenly events such as comets and supernovae thus lost their erstwhile religious significance. They could be precisely charted in Newton's clockwork or mechanical universe—a universe that afforded no qualitative distinctions between stellar and terrestrial locations.

As Newton developed his ideas, he dismantled Aristotle's theory of motion and erected new categories of thought. It no longer made sense to talk of natural and unnatural (forced) motion because these implied sentience and innate preference on the part of material bodies. One could, however, distinguish between inertial (uniform) and non-inertial (non-uniform) motion. Bodies moved inertially in the absence of outside forces; when forces impinged on them, non-inertial or accelerated motion occurred—as evidenced by changes in their states of motion. One surprising aspect of this new distinction was the equivalence of inertial motion and rest. Newton realized that since inertial motion is force-free, there is nothing to set it apart from rest. A person moving inertially could not detect that motion. She would experience nothing (no force) to suggest motion—no wind or jostling. She might, of course, see things passing by, but since every experiment performed within her reference frame yields a result exactly like that obtained from a stationary platform, she has no way of demonstrating that her motion is real rather than apparent.

But what counts as a stationary platform? Having lost the stationary earth, Newton identified a new benchmark: he equated space with rest and thereby turned space into a universal rest frame: "Absolute space, in its own nature, without relation to anything external, remains always similar and immovable."[7] So defined, space became the stationary backdrop against which the motion of objects could be precisely charted, at least in principle.

The scientific value of this backdrop can hardly be overstated. Something in the world must be absolutely fixed—otherwise distinctions between real and apparent inertial motion become impossible. But Newton's space was not quite like Aristotle's fixed earth. The earth engages the senses, and no one can deny its brute existence. Space, by contrast, is not part of our sense experience; it is, at best, the conspicuous absence of sense data. What this means is that although space may be idealized as a stationary backdrop against which to measure motion, no such measurement is possible. One cannot take space and mark it out as a coordinate system.

Newton recognized this drawback while yet holding to the proposition that "absolute space," in principle if not in practice, allows scientists to coordinate their experience with God's. Newton aspired to a God's-eye perspective, and he imagined that God's vantage point on space afforded Him faultless understanding of motion within the universe. And while it was true that space offered no sense data to the human mind, it was possible that science might detect something fully coincidental with space. If so, that substance could be exploited as a universal rest frame, thereby allowing scientists to comprehend events in a manner approximating God's comprehension.

The Universal Ether

The substance of interest was ether. Newton had proposed its existence while responding to critics disturbed by the thought of Newton's gravity propagating across empty space—that is, without a supporting medium. How could the earth hold the moon in its orbit when 240,000 miles of apparent nothingness separated the two bodies? By what physical or mechanical means would each body's gravitational force be transmitted outward? While Newton was largely content to let these questions go unanswered, believing that nature was not fully transparent to human reason, he nevertheless felt obliged to respond to those who demanded mechanical explanations for all of nature's operations.

Descartes had earlier tried to rid the cosmos of all non-contact or action-at-a-distance forces, feeling that such savored of magic and astrology, by insisting that space is a plenum: material particles fill every part of space, and these, then, mediate force or impact from one body to another. While young, Newton appreciated this approach, but he eventually concluded that Descartes populated his universe with too many imaginary mechanisms. Magnetic attraction, for example, was said to occur as threaded particles rotated into similarly threaded pores of iron or lodestone, thereby drawing the object toward the magnet. This mechanical explanation rid magnetism of long-standing occult associations, but it had no empirical support. Not wishing to "feign hypotheses" or speculate as Descartes had done, Newton stated that the deeper workings of gravity were yet "to be found out."[8] In the meantime, he would stick to incontrovertible explanations arising from the logical elucidation of empirical evidence. All the same, he did not always hold to this standard, and the ether is one of several instances where he and his followers, feeling a need to explain things mechanically, went beyond the evidence at hand.

For most, Newton's universal gravity suggested an ultra-fine substance permeating the cosmos. Vast distances separate the stars, and for gravity to interconnect the cosmos and thus function as a universal force, something other than nothingness, it would seem, had to be the means by which gravity propagated. Other ethers were eventually proposed for magnetic and electric forces, and when in the early nineteenth century light turned out to consist of waves rather than particles, the ether hypothesis became a scientific article of faith. No one could imagine waves without a supporting medium: wave energy had to be seated in a material substance. Physicists consequently were optimistic that something, however subtle, filled the apparent emptiness of space.

As the century passed, the idea of a universal ether, one that hosted all the forces of nature, became increasingly attractive. Electricity and magnetism turned out to be different aspects of the same phenomenon—electromagnetism. Light fell under this rubric as well. The outstanding exception—the force not easily related to other forces—was gravity. Nonetheless, theory called for a universal ether, and toward the end of the nineteenth century, Albert Michelson and Edward Morley devised a way to detect it. They split a beam of light in half, sending one of the half-beams on a course perpendicular to the other. Both then traveled the same distance before striking mirrors, which reflected them back to their original location and down to a detector. Would the two shafts of light meet to reconstitute the original light beam?

Theory said no. Although each shaft traveled the same round-trip distance, each moved differently with respect to the ether wind presumably caused by the earth's motion through space. One shaft would move against the wind while traveling toward the mirror and with the wind after being reflected. The other would move perpendicular to the wind in both directions. Calculation showed that the ether wind would slow the first shaft of light more than the second, thereby allowing scientists to detect the ether without directly observing it: the two shafts, upon converging, would interfere destructively with each other to a measurable extent. The

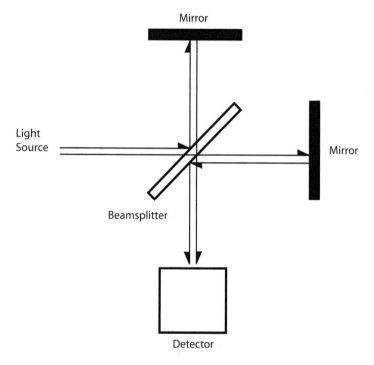

Figure 1.1. With the earth moving through the universal ether, two halves of a light beam travel the same distance perpendicular to each other. Theory indicated that one half would be slowed relative to the other (because of ether wind), but the experiment revealed no slowing.

measurement could then be used to calculate the velocity of the earth relative to the ether.

That is, relative to a universal rest frame. Newton's absolute space, physically realized as ether, would function as a stationary backdrop against which to determine the true motion of objects. It was a stirring prospect, but nature did not cooperate. The two shafts of light invariably returned at precisely the same moment, a result that challenged the ingenuity of physicists. Light, a wave phenomenon, required a material medium, just as sound waves require air. (Or, try to imagine water waves without water.) Furthermore, the speed of light had to be responsive to the ether wind, for that speed was a function of the ether's properties in the first place. And yet experiment offered no evidence of variation in the speed of light and, by implication, no evidence of ether.

Einstein's Contribution

One can begin to appreciate Einstein's accomplishment by noting that he alone broke the theoretical impasse created by Michelson and Morley's failure to detect the universal ether. While he adopted into his theory mathematical formulae developed by other physicists, his vision of reality was completely different from theirs. Unable to shake off the ether, they modified its properties to explain why it had gone undetected. Some proposed, for instance, that the ether, rather than blowing through the interferometer, blew against it so as to alter its physical dimensions. To be sure, the effect was exceedingly slight, but it was just what was needed to ensure the null result of the experiment. Simply put, the ether wind contracted the interferometer as a whole but contracted it more along the path of the forward-backward light shaft, thus making that path slightly shorter than its perpendicular counterpart. The two shafts of light thus traveled different distances at different speeds, the perpendicular shaft moving farther but faster than the forward-backward shaft. When tallied up, these differences in distance and speed neatly canceled out to guarantee that each light shaft completed its journey at precisely the same moment.

Although physicists of the era were not fully satisfied with this explanation, for more than a decade it was as compelling as any other attempt to account for the failure of the Michelson-Morley experiment. Looking back, it seems that ether theory was an attractive fiction physicists couldn't leave alone. Not only that, things got messier the more they hung on to it. In the case of the contraction hypothesis just described, one cannot help but wonder if ether had morphed into something that was, in principle and not just in practice, unobservable. By always altering the instruments used to detect it, it would remain forever undetected. Conceivably, this sort of analysis could have gone on much longer had not Einstein taken a completely different approach.

Only twenty-six years old and having failed to break into the physics profession after an undistinguished academic career, Einstein was nevertheless uniquely prepared to deal with the issue at hand. Ten years earlier while pondering the motion of light, he had devised a simple thought experiment. Imagine yourself, flashlight in hand, traveling on a light beam. What, if anything, will you see when you turn on the flashlight? Will you see a shaft of light emanating from the flashlight, just as if you were shining the flashlight in a dark and dusty cellar? Or will you see nothing at all, presumably because you are already moving at the speed of light? Keep these two answers in mind as we proceed.

Among other things, this thought experiment enabled Einstein to think more broadly than his contemporaries, some of whom were vastly more professionally accomplished. He discerned incongruities they overlooked. One involved the scientific tenet that light in an ether-filled vacuum moves at a constant velocity. This sounds Einsteinian, but the notion of light speed constancy in the late nineteenth century was anything but revolutionary. It grew out of the simple observation that the speed of waves is solely determined by the medium through which they travel. When, therefore, sound waves originate from a moving vehicle, the speed of the vehicle is not added to the speed of the waves. The atmosphere transmits those waves as if they had originated from a stationary source. Since light consisted of waves, and since its medium, the ether, was felt to be omnipresent and perfectly homogenous (its properties did not vary from place to place), it followed that light would invariably move at its measured speed of 186,000 miles per second across the ether-filled reaches of outer space.

For Einstein this kind of constancy was problematic. He realized that a beam of light, forever moving at constant speed relative to the ether, could serve as a universal yardstick against which to measure the true or absolute speed of material bodies. Imagine a spaceship moving through the solar system. Wishing to know the speed of his craft, the captain takes a speedometer reading of 25,000 miles per second. This value, he notes philosophically, is not definitive: it is the speed of the ship relative to the earth, a planet moving around the sun. Not only that, but the sun itself is orbiting about the center of the Milky Way galaxy, which in turn has its own motion arising from its gravitational connection with other galaxies. Where this hierarchy of motions ends no one knows. At this point the captain can act locally but he cannot think globally, for there is no global or universal backstop against which local motion can register itself.

But, reasoned Einstein, if (1) the ether is a substance filling every nook and cranny of the cosmos and (2) light moves at a constant speed relative to this universal substance, then one should be able to determine one's true or absolute motion without even bothering to detect the ether. All the captain has to do is measure the speed of a passing light beam. If, for example, a light beam overtakes him at a measured rate of 100,000 miles per second, he may infer a spaceship velocity of 86,000 miles per second. Should he get a reading of 186,000 miles per second, then he knows that the ship is at rest. A reading of 200,000 miles per second from an oncoming light beam (one hitting the spaceship head-on) indicates a velocity of 14,000 miles per second. This is nothing more than adding and subtracting velocities. Since, however, the velocity of light is universally invariant, any measured deviation therefrom (in a positive or negative direction) marks the true velocity of material bodies.

Einstein's analysis was ingenious, but he knew all along it was wrong. The problem went back to Newton's first law of motion. That law, as noted above, proscribes any experimental distinction between rest and inertial motion. Suppose the spaceship captain turns off the engine and lets the craft coast through empty space at a measured rate of 40,000 miles per second. Since no forces are impinging on the ship, the captain might surmise that it is moving inertially. But this he cannot prove. Should he let a ball drop to the floor, for instance, it will behave exactly as if the ship were stationary, and any other mechanical experiment (one involving material objects) will yield a similarly inconclusive result. But—and this is where Einstein's reservation about the ether begins—an optical experiment involving the constant speed of light enables scientists to make absolute distinctions between rest and inertial motion. Why, asked Einstein, would nature allow the detection of absolute motion in one realm and not the other? Why does it acquiesce to experiments involving light but not to experiments involving material bodies?

It is a mark of Einstein's genius that he could see the possibilities of contemporary physics more clearly than his colleagues. Even more remarkable, however, was his inclination to question those possibilities on philosophical grounds when he could have used them to promote consensus thinking about the ether. But his belief in an orderly, consistent universe led him to attribute the asymmetry he had uncovered to physics rather than nature. At some point, in other words, scientific thinking had gotten off track, and now optics needed to be brought back in line with mechanics. Einstein's solution was to give the speed of light not just constancy but *measured* constancy: all velocity measurements of light return the same value. With this stipulation in place, why bother to measure the speed of a passing light beam to infer one's absolute motion? Every measurement yields the same result—186,000 miles per second. Upon subtracting this velocity from the known velocity of light, our hypothetical spaceship captain gets a value of zero miles per second. He cannot say whether he is moving inertially or is at rest.

For the same reason, the observer in Einstein's thought experiment sees a flashlight beam moving away from her at the speed of light. Since she is moving inertially, anything she does will turn out just as if she were at rest. Indeed, she, like the spaceship captain, cannot make experimental distinctions between rest and inertial motion. The laws of physics do not permit such.

This return to inconclusiveness in the interest of scientific consistency came with a price tag. If something that was variable (the measured speed of light) is suddenly fixed or held fast, change must occur in new ways. Something else must give. Einstein let that something else be space and time. From now on these great fixtures of Newton's universe would be variable so as to preserve the absolute constancy of the measured speed of light.

2

SPECIAL RELATIVITY

*The irony is that the study of the phenomenon of light in
the twentieth century leads to a vision of physical reality
that is not visualizable, or which cannot be constructed in
terms of our normative seeing in everyday experience.*

Robert Nadeau and Menas Kafatos, *The Non-Local Universe:
The New Physics and Matters of the Mind*

Einstein's decision to assign a constant measured value to the speed
of light is significant in two interrelated ways. First, it sets the motion
of light apart from the motion of material bodies. The two kinds
of motion are no longer comparable. Think of a vehicle moving
inertially through space at 600 miles per hour. Even if this velocity
is constant with respect to something, it is not universally constant.
Someone overtaking the vehicle might measure its speed at 200
miles per hour. And since inertial motion is indistinguishable from
rest, this reading is just as valid (and just as inconclusive) as any other
variant measurement. What we have is a perfect democracy of differing speed values and no reason to prefer one above the others.

With light, however, the speed value is always the same, regardless of how differently its observers may be moving. Whereas it was
once thought that space and time were unaffected by the motion of

material bodies, now we may say that the speed of light is unaffected. Our speed does not alter its speed in any of the ways we are used to seeing speed altered among material bodies.

The second significant aspect of light's measured constancy is that measurement necessarily entails interaction between phenomenon and observer. In the nineteenth century physicists enjoyed the luxury of thinking of light as something independent from their investigation of light. This afforded them the privilege of regarding light as something "out there," something susceptible to measurement and analysis but fully remote from the analytic or exploratory thrust of science. Many physicists may yet defend this privilege. All the same, Einstein's innovation nudged physics toward a more experience-based view of light; that is, one from which we cannot abstract our own presence. If we insist on comparing light's motion to the motion of material bodies, that comparison immediately breaks down and leaves nothing in its wake for visualization purposes. Given what the eye and its instrumental extensions tell us about light—that indeed it has a constant measured velocity—the mind's eye cannot construct a sensible picture of light's motion. So we have only visual images gained from our interaction with light, and these do not translate into coherent visual imagery *about* light. We cannot, it seems, back away from light in order to develop a well-defined or unambiguous model of it.

No doubt this relates to the fact that in destroying the fantasy of light moving through a universal ether, Einstein did not offer a different fantasy. Or if he did, that fantasy breaks the frame of everyday reality. In his first paper on special relativity, Einstein remarked: "We shall find in what follows, that the velocity of light in our theory plays the role, physically, of an infinitely great velocity."[1] Although this role becomes clear as he continues, it is not something that lends itself to visual imagery. The mind cannot easily grasp it. Like light itself, it is expansive beyond what it illuminates.

Light-Speed Constancy

Einstein erected his special theory of relativity on two postulates. The first restated Newton's deduction regarding inertial motion: the laws of physics are the same in all inertial reference frames.

Although this was not a new idea, no one at the time appreciated it more than Einstein, for he realized that ether theory put it at risk. And to neutralize that risk he developed a second postulate: the motion of light is independent of its source.

Taken individually, each postulate reflected conventional thought. The first reached back to Galileo and Newton; the second, though hardly as venerable, reflected the contemporary attitude that light waves move through the ether at a rate whose unchanging value is a function of the unchanging properties of the ether. But Einstein had no use for the ether. Wishing to liberate physics from ether theory with its connotation of absolute space, he drew from his two postulates the striking inference that light-speed constancy is its own absolute. In some significant way, light comes before space and time, the twin absolutes of Newtonian theory.

One may imagine two inertial reference frames in relative motion. Each has a source of light—a flashlight, say—that illuminates the interior. Passengers will, of course, get the same value as they measure the speed of light radiating from each flashlight. This follows from the first postulate—the laws of physics are the same in all inertial reference frames. But what happens when light rays from one flashlight pass into the reference frame of the other? Will the two light beams register the same velocity? Given the second postulate, one must say yes. Because the motion of light is independent of the motion of its source, we cannot factor the latter motion into the former. If light is there, if it is part of our experience, it is moving at 186,000 miles per second.

To generalize, the speed of light is constant in all inertial reference frames. The import of this conclusion is far-reaching, for, as already noted, it sets light apart from material bodies, whose velocities vary from observer to observer. More fundamentally, however, it necessitates new understandings of space and time. Einstein worked these out by keeping the speed of light constant and calculating the rates at which space and time (measuring rods and clocks) fluctuate in order to preserve that constancy.

Getting Started

Two thought experiments are helpful in grasping the revolutionary aspects of Einstein's theory. The first, introduced by Einstein, undercuts the common assumption that time is a universal phenomenon. This was Newton's belief, and it yet prevails among those innocent of relativity theory. It is the view that time is a succession of universal moments, an absolutely steady march of cosmic instants—each one spanning all of reality for a "moment" and then giving way to the next. In other words, at any moment it is the same time everywhere.

The scientific value of this outlook lies in the realization that, were it true, events in the universe could be mapped onto this succession of instants in an absolutely unequivocal manner. In Chicago an economics professor just broke her piece of chalk while writing on the blackboard. This event, by Newton's lights, is simultaneous with billions of other events scattered throughout the universe. And if science were powerful enough, it could map every event onto its proper moment to recapture the sequential clarity of all reality. While this narrative might be worked out here on earth, it would transcend the limitations of any local perspective: it would be a God's-eye view of the cosmos.

Einstein's thought experiment demonstrates that no such view is possible. Time is one's local experience of time, and the notion of universal time has no basis in that experience. Figure 2.1 illustrates why this is so. Imagine an observer—let's call her Sally—who on a very dark night situates herself exactly midway between two mirrors 186,000 miles apart. (The dark night and extremely distant mirrors are just helps to the imagination; we could posit two mirrors a meter apart in broad daylight and reach the same conclusion, but then it would be harder to envisage and pick apart the different events.) She then produces a spark and waits for the light from that spark to travel the roundtrip distance to the mirrors and back to her. Given that she is equidistant from the mirrors, and that light has a constant velocity, she should see both mirror reflections at the same time. Let us call these two reflections events A and B.

We have assumed so far that Sally is at rest relative to the mirrors. Let us consider what someone—call him Jack—moving relative to Sally will see. Since Jack is moving toward one mirror (the one producing event A) and away from the other, and since, according to Einstein, the speed of light is unaffected by the motion of its source and thus is constant in all reference frames, he will see A before he sees B. This last proviso regarding the speed of light must be

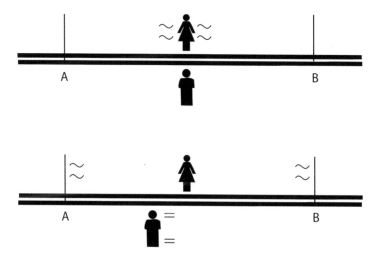

Figure 2.1. Situated midway between two mirrors and standing opposite
Jack, Sally produces a spark (*top frame*). She then sees the spark's reflections
simultaneously, but Jack, moving toward the left-hand mirror, sees that
mirror's reflection first (*bottom frame*). Since the laws of physics allow
no distinction between Sally and Jack's states of motion (Einstein's first
postulate), and since the light-mediated images do not participate in the
motion of the mirrors (Einstein's second postulate), neither sequence
of events is privileged. Both are locally, but not universally, true.

thrown in because otherwise we might find a way to vary the speed
of one reflection relative to the other so as to make the two events
simultaneous. That is, before Einstein, it would have been possible
to argue that for Jack both events are also simultaneous, since one
reflection is sped up by the moving apparatus in exactly the same
measure that the other is slowed down. But for reasons noted in the
previous chapter, Einstein could not countenance anything that
would alter the measured speed of light.

Who is right, Sally or Jack? This question, Einstein said, presupposes something that cannot be proven. In Aristotle's universe we
could have accorded primacy to Sally, at least if she were stationary
relative to the earth. Her observation, after all, would not be com-
promised by her motion. But in the modern universe the earth is
not at rest. Moreover, Newtonian mechanics allows no distinction
between stationary and inertially moving reference frames—this is
the gist of Einstein's first postulate. Jack may insist that he is at rest,

and no experiment can prove otherwise. Hence the question of how to temporally order events is answered at the local level, and there may be as many answers as there are localities.

The reader may object that the illustration shows Jack moving and that decides the issue. But for purposes of illustration we must start somewhere. Whoever creates the illustration has no choice but to imagine a scenario that no human observer can verify. If some third observer could see "what really happens," who is moving and who is at rest, she would have to see things by some agency other than light, for, among other things, it is the finite speed of light that creates the contradiction. If light moved infinitely fast then time would not lapse before events were visually announced, and all observers would agree on the sequence of events—regardless of their motion or distance from the events. As it is, however, observers cannot see events quicker than light-mediated images reach them, even though it feels like we see events at the very moment they occur.

The illustrator, then, like the nineteenth-century physicist, adopts the false posture of being removed from the system. But rigorously engaged, relativity theory disabuses us of this posture. We cannot step behind our light-articulated experience of the world to see the world "as it really is." And because we cannot get to events in some pre-light or faster-than-light fashion, light's finite velocity, coupled with Einstein's postulates, ensures the possibility of disagreement among differently situated observers.

Another Experiment

The second experiment brings us to the heart of what it means to let space and time fluctuate in order to maintain the constancy of the speed of light. Figure 2.2 shows Sally in an inertial reference frame. Time is marked out by a photon that bounces between mirrors on the floor and ceiling. To Sally, in the photon's reference frame, the photon moves vertically up and down. Every time the photon returns to the lower mirror, say, one unit of time elapses. Ideally the photon will bounce up and down indefinitely, beating out time at a uniform rate. Indeed, since the photon's speed is constant in all reference frames, this must be the case.

View from within moving reference frame

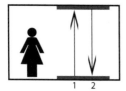

View from outside moving reference frame

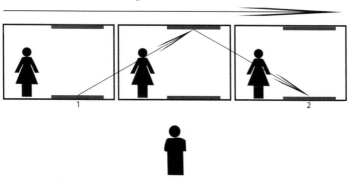

Figure 2.2. Within the compartment Sally sees the photon moving vertically; from without Jack sees it moving diagonally. Since the speed of light is the same (constant) in both situations, Jack will see the photon "taking longer" to mark out a unit of time than it would in his own reference frame.

Outside the photon's reference frame, Jack sees things differently, however. Feeling himself to be stationary, Jack sees the vehicle housing the photon clock move by. As it does, the photon travels diagonally between the mirrors. That is, as the photon bounces up and down, it must also, owing to the motion of the vehicle, move horizontally to strike the moving mirrors. A diagonal path results from this blending of vertical and horizontal motion. And because it is, of course, longer than the vertical path observed by Sally, we have another disagreement. From Jack's point of view, time passes more slowly, as it were, in Sally's reference frame than in his own. That is, Sally's units of time—as defined by the diagonal path of the photon—are "stretched out" relative to the "vertical" units that Jack would observe, were a similar clock in his reference frame.

Again, Einstein's first postulate—that the laws of physics are the same in all inertial frames—stands athwart any inclination to privilege or universalize either observation. But this is only the first postulate speaking, and without the second postulate to keep the speed of light constant, Newton's idea of absolute or universal time still lingers, despite the disagreement. This is because Jack can alter the speed of light to ensure agreement with Sally about the rate at which time flows. He need only add the speed of the moving vehicle to the photon's speed to come up with a combined velocity that allows the photon to travel a longer distance in a given period of time.

Before Einstein, this would have been the standard response to the discrepancy. Since the photon participates in the motion of its reference frame, one should add that motion to its own velocity. Sally, unable to detect any motion within her reference frame, adds nothing to the photon's speed, and so the photon moves up and down vertically at the speed of light. Jack, however, sees the photon clock moving and adds that motion to the speed of light, whereupon from his perspective the photon travels more quickly along the longer path. In both cases, the photon is calculated to reach the mirror at precisely the same moment. Thus, with photons free to move at varying speeds—free to be measured at varying speeds—time ticks away at the same rate in all places.

But for reasons explained in the last chapter, Einstein felt he could not allow such variation in the speed of light. This meant that something else had to vary, and that was time and space. Moving at the same speed in all reference frames, the photon "takes longer" to beat out a unit of time for Jack than it does for Sally. The quotation marks signify the difficulty in trying to adapt old language to a new concept. For both Sally and Jack, a unit of time (one up-and-down bounce of the photon) marks out exactly the same temporal interval, at least insofar as that interval is defined by the photon's success in hitting the mirrors. From Jack's perspective, however, that unit of time gets stretched out or dilated relative to a comparable unit of time in his own reference frame.

With a little thought, one realizes that at higher speeds time is increasingly dilated: the photon must travel ever greater distances to complete its bounce, and always at the speed of light. This at least is what Jack sees. Sally would still see the photon bouncing straight up and down; she would not experience time dilation. If, however, Jack had a photon clock in his reference frame and she were to look at it, she would make the same inferences that Jack had made about her situation. He would be the one undergoing time dilation. Since, however, neither person can prove the other to be moving, or prove oneself to be at rest, the disagreement stands. All one can say is that different localities permit different observations, and science knows no universal court of appeal.

Length Contraction

Along with time dilation, Sally and Jack would observe space or length contraction in the other's circumstance. Before seeing how this occurs, we need to state this general fact about time dilation: it happens for the observer who sees relevant measurement events in *different* places. When Sally watched the photon in her own frame, it always returned to the same place—the stationary lower (or upper) mirror. Jack, by contrast, saw it hitting the (moving) mirrors in different places, and this led him to ascribe time dilation to Sally's reference frame. Now suppose Jack wishes to measure the length of Sally's umbrella, which she holds lengthwise in her vehicle. To do this he marks a point in his own frame and notes the time required for the umbrella to fully move past that point. Then, knowing Sally's speed, he calculates the umbrella's length using the familiar relation:

$$\text{distance (or length)} = \text{speed} \times \text{time}$$

The length value Jack gets is called the umbrella's rest or proper length. For him the two critical measurement events—when first the tip and then the handle of Sally's umbrella line up with the marked point—occur in the same place. Within his own frame, in other words, time dilation is not occurring, and so the calculation yields a length value untouched by any effect of motion.

If Jack, however, attempts to somehow measure the umbrella in a single instant (both ends simultaneously), time dilation will affect his calculation. This is because this tactic entails measurement events in different places, that is, at both ends of the umbrella. Textbooks often illustrate length contraction by comparing an object's rest length with the length of the same object in varying states of relative motion. As the object moves faster relative to the observer, its length in the direction of its motion contracts or shrinks. Since objects are extended in space, seeing the whole object at a single moment involves the simultaneous witnessing of two differently located endpoints. It is this spatial separation of measurement (observation) events with regard to a single moving object that permits length contraction.

It may seem strange that length contracts while time dilates, particularly in light of the above equation, which indicates that length is proportional to time. This is where language can be misleading. The term "time dilation" may suggest more time, which in turn suggests longer time, but it is helpful to think of "longer" as referring to the longer (diagonal) light path seen by outside observers like Jack. From his point of view, not more but *less* time elapses in a given spatial interval. If in his frame the mirrors are three meters apart, the photon will traverse six meters to click off one unit of time. But looking at an identical clock in Sally's frame, Jack might see (if Sally were moving very rapidly) the photon move six meters just to reach the opposite mirror. In that case, only one-half of a unit of time will have elapsed—as defined by Sally's clock.

Once Jack, consequently, compares the two clocks (the quantity of time ticked out by each), he will conclude that Sally's clock is running slower than his own. That is, the photon in Sally's frame is taking "more time" (relative to the photon in his own frame) to complete its up-and-down bounce. This judgment is not decisive, however, because Sally would reach the opposite conclusion were she to make similar observations. In no absolute sense is time passing more slowly in either frame.

Three interrelated points should be highlighted. First, one should not mistake proper or rest length for absolute length. Proper length is merely the measurement made within the object's frame of reference. It is the measurement of someone at rest relative to the object. Since, however, that experience of rest cannot be universalized—cannot be keyed to some unmoving fixture in the cosmos—the measurement has only local validity. This fact implies the second point: space contraction and time dilation are not illusory effects. How could they be in a universe that does not allow us to privilege one observation above all others, that does not allow Jack's observations to trump Sally's or vice versa? Illusions presuppose corrective truths. The erstwhile corrective truth, in this case, was Newton's absolute space and time, but that evaporated under the glare of Einstein's analysis.

The third point was made previously but deserves reiteration. Local observations, and the realization that they are all we have, disabuse us of the conceit that we can somehow get outside of the cosmos to render universal judgments. No doubt there are ways of extrapolating from local experience to develop large and even universal explanatory schemes, but in the end we do not extrapolate ourselves from the cosmos. The very notion that we could do this assumes that we were never really threaded into the world in the first place, that we are only accidentally involved in nature. Einstein helps us to see that we are, in a deeply local way, inescapably situated in the world. This is not as dismal as it may sound, however, for light, the agency that mediates observations and thereby alerts us to our cosmic situation, is in some respects not a local phenomenon.

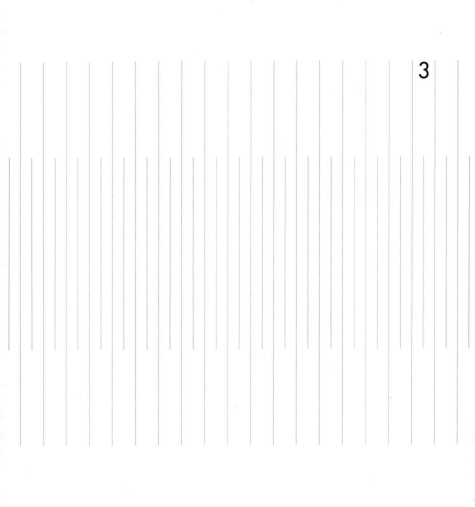

3

HORIZONAL LIGHT

So familiar are time and space that we are apt to take them for granted, forgetting that ideas of time and space are part of the shaky foundation on which is balanced the whole intricate and beautiful structure of scientific theory and philosophical thought. To tamper with those ideas is to send a shudder from one end of the structure to the other.

Banesh Hoffman, *Relativity and Its Roots*

By now the constancy of the speed of light should be well established. It should also be clear that that constancy necessitates new understandings of space and time. And this latter realization prompts consideration of the claim that light moves at 186,000 miles per second. While we are used to talking about phenomena in this way, we are not used to phenomena that undercut or relativize the very terms by which we define them. As we conventionally define it, speed presupposes space and time; it is a variable plotted against the unchanging parameters of space and time. But with Einstein we learn that space and time adjust themselves to the speed of light, fluctuate as light speed was once thought to fluctuate. Indeed this fact is built into Einstein's equations for time dilation and length contraction, where c shows up to regulate observed tempos and lengths while ensuring that material bodies do not exceed c. This is anomalous from a

Newtonian perspective: why should the speed of light show up in an equation dictating my observation of something else's length, clock tempo, and speed—something that seems completely unrelated to light? The answer to this question, I am suggesting, hinges on the term "my observation," which typically involves some sort of light-mediated visual component.

The critical mathematical factor is $\sqrt{1 - \frac{v^2}{c^2}}$, with v representing the relative velocity between object and observer and c, of course, being the speed of light. In the case of length contraction, the proper length of a rocket ship, or the length as measured by someone moving with the rocket ship, is multiplied by this factor. So:

$$\text{length (observed)} = \text{length (proper)} \times \sqrt{1 - \frac{v^2}{c^2}}$$

One can readily imagine that at everyday speeds, where v is vastly less than c, the difference between proper length and observed length will be negligible. But at speeds approaching the speed of light, the difference becomes significant. Moreover, should v equal c, the factor reduces to zero, which in turn yields zero length for a moving object being observed from another frame of reference. At this point, time dilation would also keep the next moment (the next bounce or hit of the photon in the clock apparatus) from ever occurring, so it would be as if time had stopped for the observed rocket ship.[1] None of this, of course, can happen for material bodies, but something like this *must* happen for light, at least if one chooses to treat light as a separate and separable phenomenon. Further into the book, we will question this choice, but for now let us try to answer the following question: in what sense is it appropriate to talk as if light traverses space while time passes?

An answer may be given on two interrelated levels. The first level addresses light's indifference—within its own reference frame—to space and time; the second level questions the classical assumption that events occur *in* space and time, as if space and time were distinct from those events. To be sure, we still talk of events happening in space and time, but relativity theory undermines the basis for such talk.

Light "Moving" at Light Speed

Claiming that light has an invariant velocity is not quite the same as saying that it always *moves* at the same speed. This is because light is intrinsically indifferent to space and time, the Newtonian constancies for determining speed. In *nearly* every instance one may say that light moves at *c* in a vacuum because this statement holds for all slower-than-light reference frames, even those moving at just under the speed of light. But what about light itself? In what sense does it move?

Here we are talking of light's proper motion, its motion within its own frame. By way of introduction we can say that like anything else, light is at rest relative to its own frame. To situate something in a reference frame is to make it motionless therein. Given this, we should not be surprised by the assertion that light does not move; how could it move relative to the frame that, by definition, stays abreast of its motion? But in the case of light the word "frame" can be misleading, for by suggesting spatial and temporal containment, it plays to the classical notions of absolute space and absolute time. It becomes reflexive, then, to imagine light's frame as something *within* the space-time regime, and, by implication, light is there as well.

Special relativity militates against this inclination to slip back to Newtonian modes of thought, but language can send the mind in wrong directions. Applied to any phenomena but light, "frame" is a perfectly appropriate word. This is because all slower-than-light phenomena are spatially and temporally bounded or framed, even when time dilation and length contraction are ascribed to them. But at the speed of light, those bounds fall away: in the mathematics of special relativity, space (length) and time collapse to zero magnitude. They are no longer there to frame light.

Part of my purpose is to propose that there is a sense in which light is unframed and that this feature of light shows up not only in special relativity but also in the unbounded seeing experience. But before moving toward this idea, we need to say more about light's relation to space and time. Implicit in what we have said already is the notion that light is indifferent to space and time, that it is not reducible to "matter or its motions."[2] Some thinkers have tried to grasp light from within its own economy and have variously concluded that it is ageless or atemporal, or that "[l]ight and influences propagated by light make zero-interval linkages between events near and far."[3] Here is what Sydney Perkowitz says about light: "To the best understanding we can muster . . . the universe is made so that light always travels its own distance of zero, while to us its clock is stopped and its speed is absolutely fixed. These sober conclusions read as if they come out of some fevered fantasy. Light, indeed, is different from anything else we know."[4]

As counterintuitive as these judgments may be, they follow from special relativity. That theory, far from collapsing the mystery of light, raises it to a new level. Not only do we fail to understand light's motion (speed), we are hard-pressed to grasp a phenomenon that is intrinsically indifferent to space and time. The two difficulties, of course, go hand in hand. Embedded in space and time, our bodies spatially and temporally configured, we have little capacity to imagine the universe from light's point of view. Nevertheless, it is worth trying. In noting that the "solid, stable world of matter appears to be sustained at every instant by an underlying sea of quantum light," Bernhard Haisch remarks:

> If it is the underlying realm of light that is the fundamental reality propping up our physical universe, let us ask how the universe of space and time would appear from the perspective of a beam of light. The laws of relativity are clear on this point. If you could ride a beam of light as an observer, all of space would shrink to a point, and all of time would collapse to an instant. In the reference frame of light, there is no space and time. If we look up at the Andromeda galaxy in the night sky, we see light that from our point of view took 2 million years to traverse that vast distance of space. But to a beam of light radiating from some star in the Andromeda galaxy, the transmission from its point of origin to our eye was instantaneous.[5]

Of course no one can ride a beam of light as an observer, so no one can witness what Haisch describes. All the same, it is remarkable that there is something in the cosmos that intrinsically collapses the universe to a dimensionless point and thereby upsets our notion of the cosmos as an all-inclusive receptacle. This fact points us toward the question of what it means to be *in* the cosmos.

Being in the Universe

There is in science, writes Stanley Jaki, an "old insensitivity" that deems the universe typologically similar to objects found in the universe.[6] Jaki makes this point because he is doubtful that scientists, unable to extricate themselves from the universe, can fully understand it. As scientific understanding has been idealized in the West, the prospect of perfect understanding presupposes apartness or objective distance, an "immaculate perception" borne of non-participation. But how, asks Jaki, would scientists de-couple themselves from the cosmos?

Jaki's point informs the issue at hand. More so than other branches of science, modern physics reminds us that we are threaded into nature, that nature and human knowing are mutually constitutive at the level of local experience. A case in point is special relativity. Picking up the other end of the stick, the youthful Einstein thought about local experience while questioning such global abstractions as universal ether and absolute space. Inclined to accept only what fell under the observation of the senses, he developed a theory that coordinates fact or knowledge with local perspective. Others, of course, had noticed that the world lines up differently for different observers, but Einstein mined this experiential truism more deeply than his predecessors. He brought local experience back to the forefront of physics, a move that not only advanced science but also suggested that science sometimes advances by returning to its roots in lived experience. In that experience, observer and object do not always neatly peel away from each other, and this seems to be the case in special relativity, where an object's length, say, is constituted by the compound system of object and observer.

Length, one may say, is observer-mediated, and the agency of mediation is light. Of course light is the means by which we visually apprehend an object's length, but its speed is also a background constancy against which length shows up as a *relative* value—a value specifying the object's relation to a particular observer. When grasping this point, it is important to resist the classical assumption that every object intrinsically possesses a standard or absolute (privileged) length, which then, under varying observational circumstances, takes on different appearances. For material bodies, there are no such standard or default values; there are only observed or measured values, and observation implies linkage or relation with observers. Thus, when it comes to properties such as length, mass, and time, one may not, in the interest of getting in touch with reality, fall back on observer-independent values. To follow Peter Kosso: "Things *have* length only with respect to a specific reference frame. This complication, this relativity, is not in how you look at things; it is in how things are."[7] The different appearances or observations go all the way down and thereby constitute reality: there is nothing to belie or correct them at a deeper level.

Instead the corrective is found in the dual realization that (1) nature and the knowing thereof is a bilateral, mutually formative relation, and (2) absoluteness resides in that relation—not somewhere in the outside world. Being a *measured* or *observed* constancy or absoluteness, the speed of light belongs as much to the mind and our perceptual apparatus as it does to nature. It denotes a point at which mind and matter—epistemology and ontology—achieve mutual immanence. This is why some have found it useful to think of light, or the speed of light, as a horizon, something caught at the interface of physical and perceptual reality.[8] Not fully coincidental with the physical features or places that set them off, horizons recede with our advance, thereby neutralizing attempts to overtake them. This, of course, is because they are complicit with the act of seeing them. Our movement, our changing angle of vision, causes the horizon to shift or retreat, and so it is too much a part of the seeing experience to be marked out in space and time (as, for example, at a specific place on the earth) for independent viewing.[9]

In a similar manner, light (which helps to fashion horizons) appears to be too complicit with the seeing experience to be marked out for independent viewing. Like horizons, light is indifferent to the speed at which we attempt to overtake it—thus the constancy of the speed of light. It is interesting to note also—and this recalls Einstein's observation that in his theory light plays the role of an "infinitely great velocity"—that horizons suggest the infinite while demarcating the finite. Absolute invariance in the face of local change —the experience of stepping off toward the horizon without closing the interval—prompts thoughts of infinity. Some desert Bedouins instinctively assume that life goes on forever because, travel as they may, they never reach the horizon.[10]

Here is another way of making this very fundamental point. Ask yourself if you ever see light per se. No doubt you see lighted objects, but do you see light in the absence of objects? Stephen

Palmer, a prominent vision scientist, writes that "visual perception implies . . . three things: light, surfaces that reflect light, and the visual system of an observer that can detect light." He then states: "Remove any one of these ingredients, and visual perception of the environment simply does not occur."[11] Put differently, light and eye are not sufficient for the experience of light. That experience registers only when material surfaces are thrown into the mix. So while it may be convenient to think of light as something apart from the experience of light, there is no empirical basis for such thinking. Light shows up or announces itself only at the interface of physical and perceptual reality. That no doubt is why light moves as horizons move—with our motion in mind, as it were, because our mindful awareness of the world is already implicated in the world's reality.

Horizonal Constancy

On a perfectly spherical planet, the horizon, showing up at the interface of physical geometry and visual experience, would remain a constant interval ahead of us—no matter our speed. The analogy is doubly apropos because modern cosmologists sometimes liken the geometry of spacetime to the surface of a sphere, albeit with additional dimensions. Contrary to everyday thought, then, we are not *in* the universe as an apple is in a bushel basket. That is, the word "in," as conventionally understood, does not capture the geometric character of the universe: it erroneously suggests a bounded cosmos with a unique center and outer edges. In fact the universe, somewhat like the earth's surface, is intrinsically unbounded and indifferent to center-edge distinctions. Or if such distinctions are made, they are relative to the person making them. Center and outer edge shift as perspective shifts, just as on the sphere of the earth, one's experiential center and horizon constantly shift. In neither case does it make sense to talk about being in a closed system, a system that obviously bounds our experience and thereby contains us. This, I believe, is because nature's bounds or limitations inform our own nature. They do not, consequently, loom up as *extrinsic* fixtures curtailing and confining our experience of the world.

We can press this thought to deeper import by pointing out a defect in the analogy upon which it is based. When viewing horizons on the earth, we feel apart from the earth, notwithstanding gravitational and organic (evolutionary) tethers. That no doubt is because we know of non-terrestrial locations—planets and stars—and we can readily imagine being on these bodies and seeing other horizons. Each horizon would have a different constancy about it, depending on the size of the star or planet. Thus, we might reasonably conclude, there is nothing absolute or universal about any horizonal constancy, the earth's included. But—and here comes the disanalogy—if the universe similarly has a particular horizonal constancy, would we be able to apprehend it as a relative constancy? Could we find another platform, another universe, from which to see things differently? If Jaki is right, no. We cannot migrate out of the cosmos to achieve an outside perspective.

What is more, any pretense of stepping out of the cosmos is foiled by the modern realization that even at the local level (the level of individual objects) space and physical matter are mutually constitutive. This is what Milič Čapek has in mind when he speaks of the fusing of space "with its dynamic and changing physical content."[12] Or to follow John Wheeler: "[S]pacetime geometry is not a God-given perfection standing high above the battles of matter and energy, but is itself a participant in the world of physics."[13] More prosaically, different words—space and matter—do not denote separate or separable realities. Material bodies could no more fall out space (or, more precisely, spacetime) than a rock could fall out of the atoms that compose it. If something, therefore, is said to be *in* the universe, it is not as a loose part rattling around in a bucket. Rather it is implicit in the cosmic structure.

Quantum Considerations

These or similar points are routinely made in the register of quantum theory, and often the intent is to undermine the Cartesian assumption that mind is distinct from the material world. Physical substance, said Descartes, does not partake of mental substance because the former has no capacity for psychic activity. Not only that, but mind, unlike matter, is said to occupy no space or place and undergo no motion. This discontinuity between mind and matter has prompted much discussion, and many have found reason to reject it. Nevertheless, it seems hard to eradicate from our thinking, despite pronouncements such as "The laws [of physics] leave a place for mind in the description of every molecule."[14] Werner Heisenberg, one of the architects of twentieth-century physics, lodged this complaint against Cartesian dualism: "[T]he common division of the world into subject and object, inner world and outer world, body and soul, is no longer adequate and leads us into difficulties."[15]

Chief among the difficulties is our inability to sustain the dualism in the face of events where mind, by seeming to bury itself in matter at the quantum level, overleaps its Cartesian function of merely taking cognizance of matter. The knowing subject blends into the known object, and blends in so intimately that we cannot tease the two apart. As a consequence, any solution to or understanding of such events rests on the prior understanding that mind and matter are in a state of solution, at least to some small degree that nonetheless overturns the Cartesian dream of knowing reality down to its most minute detail.

In modern physics, uncertainty or indeterminancy resides in the minute details, because there Heisenberg's uncertainty principle shows up most dramatically. One explanation for this uncertainty posits an interference effect between the world we see and the light by which we see it. Ordinarily this effect goes unnoticed, for impinging photons only negligibly disturb large objects like tables and chairs. But at the subatomic level of reality, particles like electrons are readily knocked about by energetic photons. And since those photons—when they bounce back to our eyes or against scientific instruments of detection—are the means by which we bring the world into focus, we fail to "see" subatomic particles with perfect clarity. The agency of revelation—light—is subtle, but not so subtle that it does not disturb and thereby obscure the world it illuminates.

What this means is that we cannot see the fine detail of physical reality for light. This, at least, was the early interpretation of the uncertainty principle. The mature and more radical interpretation— the Copenhagen Interpretation—asserts that subatomic particles are intrinsically imprecise. When left alone (unmeasured), they do not possess precise, particle-like position and momentum values. Instead, they are probabilistically spread over regions of space and time incorporating many such values. Precision comes with observation: when measured for position, say, probabilistic entities—waves—collapse to a single value.

In the Copenhagen Interpretation, then, indeterminancy comes first—is a kind of ground state—and nature is responsive, within probabilistic limits, to acts of observation. Those acts do not change precise properties, but rather bring them forward from an ensemble or superposition of possibilities. One inference that may be drawn from this outlook is that mind and matter are already in a state of solution, and a conscious act on the part of mind—an observation—is required to effect seeming separation or duality. But the observational posture, while trading on the assumption of mind-matter separability, does not actually bring it about. That is why, to cite Heisenberg again, the subject-object distinction "leads us into difficulties." Not there in the first place, the distinction or duality can never cleanly take effect, and so at the quantum level we find messy remainders—or, to speak more literally, re*mind*ers—of mind's deep immersion in nature. Quantum reality, however, is not the only place that such reminders show up. We find them also in relativity theory and, most surprisingly, in everyday experience.

This happens, I submit, because the speed of light specifies a relation between observer and things observed. It is a constant of nature—no one doubts this. But that constancy, I am proposing, reaches into our knowing of nature. More than a physical fact, it is part of the way we apprehend physical facts. The constancy denotes mind-matter interblending. This is why the speed of light "is really not like other speeds."[16] It is not, for example, like the speed of a baseball. While a baseball is clearly within the spacetime regime, the same cannot be said of light. To be sure, light shows up in the spacetime regime as it interacts with material bodies, but its speed is the escape velocity thereof.

Light thus comes off as both this-worldly and other-worldly. Even as it shows up in our experience, it is much harder to pin down than something like a baseball. A baseball is spatially bounded; light is expansive. One can hide a baseball (at a definite location in spacetime) and draw a map of how to find it. Light, by contrast, does not lend itself so readily to being cut off from other features of reality; it must be mapped differently. Finally, baseballs wear out and fall apart but special relativity prompts the suggestion that light is ageless or atemporal. Clearly light invites further study.

4

EXPERIENTIAL LIGHT

Behold the light emitted from the Sun,
What more familiar, and what more unknown?
While by its spreading Radiance it reveals
All Nature's Face, it still itself conceals.

Richard Blackmore, *Creation: A Philosophical Poem*
(London, 1712)

When two colors are perfectly blended together, constancy or homogeneity replaces difference or heterogeneity. Something like that appears to occur with light speed constancy. Light is such that we cannot see it without seeing *by* it, a fact that keeps us from separating light from our experience of light—the two are coincidental. Of course, it is instinctive to treat light as if it were distinct from the visual experience, but this attitude sets the stage for difficulties. One difficulty is the puzzling fact that our own speed or motion fails to affect the speed of light. Assuming that these two speeds are unrelated, we are surprised to learn that no motion or maneuver on our part can alter the speed of light. What we have overlooked is that by light we are visually blended into the world. Occurring before we undertake measurements of the speed of light, that blending pre-decides the question of light speed in favor of a homogenous or universal value.

In this chapter I try to develop this point by showing how two puzzles in physics may be clarified by treating light as an experience rather than an abstract (remote-from-experience) phenomenon. First, however, it is important to reiterate that, even in science, light never slips the tether of experience. "When we investigate a natural process such as light," writes Eduard Ruechardt, "we never observe the isolated process as such. We always use some instrument—lenses, prisms, optical gratings, interferometers, or photographic plates or photoelectric cells—but in each and every instance we use at least our eyes."[1] To restate in terms central to my argument: (1) light is never seen per se, but always in conjunction with other things; and (2) even when constrained by instruments, light inevitably loops back as experience. Ruechardt mentions visual experience in this regard, and certainly there is no better instance of light's reality. Later in the book, I tentatively propose that light also reaches into other realms of experience. For now, however, it suffices to attend to visual experience.

Light Speed Constancy in All Reference Frames

The first puzzle is the one we started with, and I have already tried to suggest a solution. But there is need for elaboration. Einstein's second postulate is sometimes given as the constancy of the speed of light in all reference frames. Here "all" implies frames moving at any and every possible speed. The puzzle, of course, is this: why should the speed of light be constant when some frames are moving faster than others?

This indifference to the motion of reference frames sets light apart from those frames: their measured speeds may vary from one observer to the next, but the speed of light holds still for all observers. It is fast in the double and seemingly contradictory sense that it is very fast—the fastest possible velocity—and an absolutely fixed value. No other phenomenon manifests itself so strangely, and that strangeness constitutes the uniqueness of the speed of light.

But this uniqueness may be traced back to another uniqueness relating to light and vision: light, unlike light-illuminated objects, is its own messenger, and so "nothing, not even light itself, can bring us news of its upcoming arrival"[2] This is John Schumacher's point. He insists that while light visually announces faraway objects, it cannot announce itself as a distant phenomenon. As the ultimate messenger—there is no swifter signal in the cosmos—light signals its arrival only when it arrives, not a moment sooner. Put another way, we never see light in intermediate space. It dawns only in the moment of immediate experience—when photons strike the eye. Indeed, the very idea of seeing light off in the distance is incoherent. By what means would we see it, if light, the agency of seeing, were removed from us?

What this implies is that no measurement or observation of light is ever realized in an objective fashion. Never is light per se experienced beyond one's location or reference frame. And, to follow the logic of physics, phenomena *in* reference frames register similar or constant speed values with respect to similarly situated observers. In the case of material bodies that value is zero velocity—they do not move relative to their own frames. In the case of light, which is not fully reducible to "matter or its motions,"[3] the value is 186,000 miles per second. In either case, there is a constant measured value for the simple reason that observer and phenomena are similarly situated.

Accustomed as we are, though, of thinking of light as something "out there," as something we can see per se in other frames, we find light speed constancy puzzling. Light may well be in other frames, but only as a matter of visual experience for all observers, and while light allows us to see other reference frames, the seeing experience invariably originates in our own. When light strikes the eye, our own visual experience occurs, and we are at the very center of it. That spark of vision—photons impinging on the retinas—is always a local event, always an event centrally located in *this* reference frame.

This is not idle conjecture. The constancy of the speed of light compels the realization that light centers every observer—even when events observed are far away. Imagine Sally producing a spark. Given light speed constancy, light from the spark will radiate away from her in all directions at a measured rate of 186,000 miles per second. Thus she is at the very center of a circumambient light sphere. Now imagine Jack hurtling away from Sally at, say, half the speed of light. When he sees the spark, its light will *also* expand away from him in all directions at a measured rate of *c*. Thus he *also* will be at the very center of the spark-initiated light sphere, even though he is far removed from the spark and far removed from Sally. And so it goes for every person who sees the spark. Visually speaking, light speed constancy puts every observer right at the middle of the world, even when observers watching the same event are far apart.

This is the deeper meaning of the relativistic proposition that there are no privileged reference frames. When occupied, each vantage point becomes a center, a place from which the world expansively and isotropically coheres according to the laws of physics I know to be operative at *my* location. My experience, or my experiment, therefore, has universal significance, even though it is local rather than universal. It is as if the world pivots around any and all points, so long as those points are occupied by observers.

If this analysis seems counterintuitive, it is because while visual experience begins in one's own reference frame, it does not end there. Hans Jonas notes: "[S]ight is *par excellence* the sense of the simultaneous or the coordinated, and thereby of the extensive. A view comprehends many things juxtaposed, as co-existent parts of one field of vision. It does so in an instant: as in a flash one glance, an opening of the eyes, discloses a world of co-present qualities spread out in space, ranging in depth, continuing into indefinite distance."[4] This wide visual expanse misleads us into believing that light—like the distant objects populating the expanse—is available for objective seeing. And given this erroneous belief, we reflexively compare light to those objects. We expect that its velocity will vary as we alter our own. Since, however, the light-originated visual experience originates with us (as light strikes the eyes), light is already in our reference frame and fully implicated in our experience. It cannot, consequently, be measured at relative or variant velocities, the very thought of which presupposes apartness.

This, I believe, is a beginning to the story of light speed constancy. To sum up, light is so close to home, so deeply implicated in our "reference frame" and implicit in our nature, that we cannot step away from it for objective viewing. At one level it informs us of the world, but at a deeper, more fundamental level it informs or structures the world we see and the very way we see it. It is not just the agency of sight but a universal precondition thereof. Jacques Derrida had something like this in mind when he asked, "Who will ever dominate it [light], who will ever pronounce its meaning without first being pronounced by it."[5]

Another Puzzle: Nonseparability and the Hereness of Light

The second puzzle is related to the first. Trying to explain light speed constancy, I characterized light as something inevitably in one's own reference frame. But this characterization is not quite right, for it suggests that light is like the material objects that make up and occupy a reference frame. And yet, as noted above, light expansively lights up other reference frames. Unlike material objects, it cannot be easily assigned or confined to particular locations.

Here, however, I am thinking only of the experience of light—the way light visually stretches us outward. Ignoring experience, let us think about parts or particles of light. These are called photons or light quanta. Max Planck proposed their existence in 1900, and they have since become a standard fixture of physics. Not surprisingly, they encourage the thought that light may be pinned down to particular locations: the photons striking my eye are not the ones striking your eye. This thought or assumption, however, has been challenged by experiments demonstrating the nonseparable character of photons. That is, under certain conditions, different parts or particles of light defy the expectation of apartness or separability.

One gets a sense of the uncanny nature of photons by noting that they are not easy to characterize as discrete entities. When first conceptualized, they were thought of as atoms or pellets of light: self-contained and precisely located in space and time. Subsequent research has shattered this characterization. George Greenstein and Arthur Zajonc point out that it is now difficult to think of photons as countable units:

> Many people tend to think of photons as being the constituents of light, in the same sense that atoms are the constituents of matter. But this view is erroneous. While a brick contains a

definite number of atoms, the same cannot always be said
for light. Most forms of light—sunlight, for example, or light
from electric bulbs—are not composed of one photon, or ten
photons, or any particular number of photons at all. . . . The
central lesson . . . is that the concept of the photon is far more
subtle than had been previously thought.[6]

To count photons, one needs separable entities, but experiments
designed to determine whether (supposedly) spatially separated photons are separable indicate otherwise. Put differently, the photons
seem to remain instantaneously connected or entangled while moving apart. Not only that, but the connection is not attenuated or de-
layed as distance increases. It seems indifferent to space and time.

At first glance this result appears to challenge Einstein's claim
that no signal or causal influence can exceed the speed of light. But
it is not clear that nonseparability entails the transmission of signals
or causal effects. If that were true, it could be exploited for informational purposes, and for several reasons this seems impossible. For
one thing, sending messages assumes manipulating phenomena
according to some prearranged code: one lantern if by land, two if
by sea, for instance. But nonseparability is already in place before
such manipulation can occur. This, I believe, is because manipulation as we know it always amounts to manipulation from without,
where we the manipulators are apart from things manipulated. The
problem is that at the quantum level nature does not let us get away
with this pretense.

Fully implicated in the system, we cannot exploit it to our advantage.

My intent is not to explore the details of nonseparability, only to
note that it is, from the point of view of classical physics, very puzzling. How can two counter-propagating photons remain in timeless
contact? Given the premise that light can be broken into different
parts, this is a perfectly natural question to ask. But perhaps that
premise is wrong. We have already seen that photons—the so-called
smallest parts of light—do not adhere to our conventional understanding of separate, countable parts. I have suggested, moreover,
that light as visual experience does not lend itself to separability and
particulate division. We may talk as if it does, but such talk often
covers over issues related to nonseparability.

For example, one may talk with the greatest of ease about photons striking the eyes and initiating visual experience of distant objects. This makes vision sound like a two-event affair. But in the moment of vision we do not experience two events: we simply see. That is, photons per se are eventless. They do not register themselves apart from the visual data they convey. For all we know from experience, they are those data. Furthermore, if we choose to think of photons as tiny, local, and particulate or atomistic, then it is odd that the visual experience they initiate is expansive, distant, and seamless. That experience hangs together in a large, unitary way, while photons come off as discrete and pluralistic. At least they did until their nonseparable nature was experimentally detected.

I hold that light, in giving us visual experience, already behaves as a principle of nonseparability.[7] Light is *here* striking the eye but offering immediate witness of something *there*. Before elaborating this argument, let me indicate the main points:

1. By means of light, material bodies offer images or reflections of themselves.

2. This allows them, as in the case of mutually reflecting objects, to interact at a distance.

3. Light, however, is constitutionally unable to offer an image of itself, for the clear seeing of material bodies necessitates a clearness or clarity on the part of light that amounts to invisibility.

4. Lacking the ability to offer an image of itself, light also lacks the ability to interact at a distance by means of images.

5. If, therefore, interaction occurs between separate parts of light—individual photons, say—it does so on a basis different from that of material bodies.

6. This basis is already given by visual experience. As the swiftest signal in the universe, light is its own messenger, and nothing is seen until light arrives. Or, to express the point differently, images of light cannot race ahead to announce light's upcoming arrival, for, given light's nature, such cannot exist.

7. In brief, light is a principle of arrival or immediate presence. Always occurring *here*, it cannot interact with itself across different spacetime locations.

8. Lacking the means to interact at a distance, light must interact at the same place, which suggests that within each part of light (each photon) all other interacting parts are wholly present or fully arrived.

9. Proposition 8, though hard to fathom, is consistent with attributions of light's nonseparability in modern physics.

The argument begins with the thought that light is different from material bodies in that it cannot offer a facsimile or image of itself. That is, it cannot split itself into here and there by means of a projected image. This unique attribute or inability stems from the fact that unimpeded projection of image requires an imageless medium, a messenger that does not announce itself in the process of announcing other things. To bring us clear or distinct images of other things, light itself must be clear, albeit in the other sense of the word: invisible or imageless. If it were visible in its own right, it would, in a manner alien to its nature, obscure our view of those things.

Perhaps the assonance of *invisibility* and *indivisibility* is not accidental.[8] My point is that each implies the other. We cannot see light qua light because it is, from an experiential point of view, inseparable or indivisible from—identical with—the images it conveys. Seeing of course is a matter of *vision*, but it is also a kind of light-based *division* whereby objects show up in places they have no material existence—namely, in our minds. Light, in other words, gives us the here-there split: the perceiving mind is not co-located with the perceived object. But that split is not intrinsic to light, which, if it is to do its work of vision clearly and cleanly, must lack the ability to cast off images of itself. So, light remains invisible or imageless and at the same time fully intact or indivisible.

To come at the point from a slightly different angle, by means of light-mediated images we are able to have visual experience of things not materially present. As a presenting medium, however, light does not yield to such experience. To bring off vision of other things, light must be "the 'letting-appear' that does not itself appear."[9] It must also be the "letting appear" that gives rise to visual experiences not spatially and temporally coincidental with the things experienced. In letting all this appear or happen, light is not party to any of it. Its lack of appearance, its clearness or invisibility, keeps it fully present rather than imagistically divided off into places we call "there" and "then."

Put another way, the terms or conditions of vision are light's invisibility *and* light's presence or hereness. The first condition (invisibility) has long been acknowledged, and it implies the second condition (presence), for light is light *only* when it is here initiating visual experience, never when it is remote from us. But light's presence has escaped comment, save for the observation that light cannot be other than *here* when it announces the existence of things not here.[10] I am proposing that light's presence or hereness is absolute, and that this absoluteness is consistent with claims of nonseparability in modern physics.

A thought experiment will throw these ideas into sharper relief. Thinkers sometimes compare the interactive vastness or nonseparability of photons to Indra's heavenly net of pearls. Each shiny pearl, though distinct from the rest, gathers in the reflections of all other pearls and so becomes a microcosmic recapitulation of the entire net. This analogy helps, but it becomes a disanalogy when we consider that photons, unlike pearls, are pure light. Mutually reflecting pearls give us infinity (a never-ending series of images), just as mutually reflecting mirrors do.[11] Reflections, however, are light-mediated images of opaque or semi-opaque objects, and being what it is, light cannot project or mediate an image of itself. If it arrives at all, it must arrive in person, as it were, not as a facsimile or representative image. Thus, the common involvement of photons is different from that of mutually reflecting pearls. Whereas pearls merely gather in reflections of themselves, photons go the full distance and gather in each other. And so, if there is interaction among different parts of light, there must also be undifferentiated wholeness, for each part of light is gathered into all other parts.

This holistic interpenetration, which would seem to deepen forever in the manner of mutually reflecting mirror images, gives us a basis for understanding why light manifests itself in ways that bespeak nonseparability. Not able to project images of themselves, photons cannot interact *at a distance* as pearls or mirrors do. If they interact, that interaction must occur at the same place, the *here*, which necessarily makes their interaction interblending or integration.

Admittedly, this argument is counterintuitive, for it is reflexive to class light with light-illuminated bodies. But when scrutinized, light resists such classification and often breaks the frame of everyday reality. Here is what physicist Arthur Zajonc, chastened by his long study of light, concludes regarding our natural inclination to break light into parts or pieces:

> Try though we may to split light into fundamental atomic pieces, it remains whole to the end. Our very notion of what it means to be elementary is challenged. Until now we have equated smallest with most fundamental. Perhaps for light, at least, the most fundamental feature is not to be found in smallness, but rather in wholeness, its incorrigible capacity to be one and many, particle and wave, a single thing with the universe inside.[12]

Some resist such analysis by asking why light should be singled out so uniquely and dramatically. My response is twofold: (1) light cannot be singled out in any absolute way; its very nature is to remain nonseparable or relational; and (2) something about our world secures the possibility of asking questions about it. That something, that coherence or intelligibility, precedes and fosters the questions we ask and therefore functions as a kind of primordial given that cannot be fully grasped. One very venerable stream of thought identifies light as this fostering agency. This point is developed in the next chapter.

5

RELATIONAL LIGHT

Inasmuch as God sprinkled His light upon
humanity, human beings are essentially one.
In reality, His light never separated.

Jalal Al-Din Rumi, *Rumi: A Spiritual Treasury*

T. S. Eliot once wrote of "a music heard so deeply that it is not heard at all."[1] I am saying something similar about light. Light, I believe, is so deeply informative of our nature that we cannot bring it forward in a detached, informational way. Expressed differently, as the very coin of illumination and understanding, light cannot be traded against itself. Hence, we will never overtake it as something apart from us. That would be like overtaking the horizon or flying skyward to get a closer look at a rainbow. Because horizons and rainbows do not precede visual experience, they cannot be detached therefrom and explained as freestanding phenomena, and so complete explanations always reference observers. In a more fundamental sense, the same is true of light. Any earnest explanation of light must always reference observers. This is because light, while informative of the world, is simultaneously informative or constitutive of our own nature. Going both directions at once, light informs or shapes our knowing faculties while informing us of the world.

Twenty-five centuries ago Heraclitus seems to have grasped this point. For him *logos* was the primal cosmic principle, manifesting itself as "the brilliant fiery stuff which fills the shining sky and surrounds the world."[2] He let the manifest brilliance of the world command his thinking and concluded that the shining of nature is the showing of truth. Or, to put the matter more simply, for truth to be *shown*, it must be *shone*: truth implies brilliance. Furthermore, Heraclitus (much like Plato after him) argued that human understanding arises from the consonance that exists between the outer *logos* ("world-fire") and one's inner *logos* ("soul-fire").[3] Philosophy or the pursuit of wisdom could never be the result of mere curiosity, for curiosity would be impossible if there were no preexisting harmony or isomorphism between mind and world. That harmony enables understanding by providing the mind with targetable phenomena; the lines of harmony, as it were, undo the mind's self-containment. Better said, they never really let that self-containment occur in the first place.

What Heraclitus wanted to know was how understanding of the world is possible *in the first place*? What sparks the wondering, thinking, knowing process? For him that spark was not hard to find: it was all about him, the fire-illuminated world, or the world made intelligible by fire. The match that struck the world into fiery existence simultaneously struck a fire in his own mind. *Logos* was thus the first place or thing, the primal principle of intelligence or light that grounds and enables human understanding.

Read this way, Heraclitus collapses the distinction between mind and world that so deeply shapes modern thought. Still, the skeptic is inclined to ask what makes light so extraordinary? Why single it out as if it were unique? This is a fair question and one very much in the spirit of modern science, which since the seventeenth century has broken down Aristotelian distinctions between the earth and the starry heavens (the sublunar and supralunar spheres). With this success, the universe has become homogenous, a system thoroughly knit together by common laws and substances. (No longer does one approach the utterly transcendent reality of God, as Dante did in his *Divine Comedy*, by voyaging toward the stars.) The outstanding exception to this rule of commonality, however, is mind. Modern science has traded on the Cartesian assumption that mind is not knit into the cosmic system; observers consequently can view reality without participating in it. So another fair question, it seems to me, is what makes mind so special?

In what follows I propose that light, while extraordinary, is also deeply ordinary owing to its relationality. Einstein suggested that light, while possessing a finite velocity, "plays the role, physically, of an infinitely great velocity" by virtue of its constant measured speed.[4] That speed, I have argued, specifies a universal relation between observer and things observed. The eye sees things, but only as light brings the two into relation, and the objects seen—their spacetime properties—fluctuate so as to preserve the constancy of the speed of light. This relationality goes all the way down: never does light show up as something apart from the things we see. So the extraordinary essence of light consists not in self-announcement but rather in its ability to remain completely relational: its ability to bring "ordinary" things into mutual relation while absenting itself as a thing per se.

Light Then and Now

As just noted, the modern inclination is to treat light as another familiar entity. Newton ascribed to it material and mechanical properties. Its fundamental components—what Newton called *corpuscles*—obeyed his three laws of motion and were thus, in principle, fully within the reach of human understanding. This attitude contrasted sharply with reverential views of light that marked premodern thought. Dante ends his *Divine Comedy* by paying homage to eternal light. "In its profundity," he says, "I saw—ingathered and bound by love into one single volume—what, in the universe, seems separate, scattered."[5] So densely packed with reality was this light that a moment's contemplation thereof weighed more heavily on Dante, and slipped more easily from his memory and understanding, than twenty-five centuries of recorded history. But despite his inability to hold on to the vision, Dante came away knowing that eternal light embraces the miracle of harmonizing the upper and lower worlds. Therein two seemingly incommensurable magnitudes

—divine perfection and human imperfection—are brought into relation.

This is just one example of light's universal ability to inspire awe and metaphysical reflection. Like Robert Grosseteste, another medieval investigator of light, Dante distinguished eternal light from physical light.[6] Still, the latter dimly approximated the former and thereby functioned as a portal to divinity. Similar thinking—that the luminosity of the physical world signifies yet another, higher reality—shows up consistently in ancient cultures. By "chant[ing] the radiance" of the world,[7] Vedic hymns acknowledged unseen agents or agencies whose "light touch" yet hung in the air. For early Mesopotamians, the radiant sky was a divine and overpowering presence, apparently "the very source and center of all majesty."[8] Hard at first to push away or objectify, the sky gradually became the "mere abode" of deity rather than deity itself.[9] In ancient Egypt, the gods were the "immortal shining ones" who rained down intelligence as they made their daily (or nightly) trek across the sky.[10] This outlook, though somewhat muted, flourished among the Greeks: the sun, moon, and stars—our primary sources of light—partake of divinity in their mirroring of eternal principles and their eternal, incorruptible nature.

One would be hard-pressed, I think, to find a pre-modern culture indifferent to light and its evident ability to grace the world with a suggestion of divine presence. For all that, modern culture as a rule does not accord special status to light. The exception to this rule, perhaps, is found in modern physics, where light has turned out to be a very puzzling phenomenon. And yet most physicists hardly feel a need to adopt the old reverential language when talking about light. Granted, light may be puzzling, but it is not a spiritual mystery. Or if it is, physical light—the light that is all about us and that which science targets—is something else altogether, and spiritual light is merely a religious fancy.

Consider, however, that both modern cosmology and religious literature regard light as a first principle or primal reality. In the Judeo-Christian tradition God calls forth light before implementing the physical creation. Similarly, the big bang—modern science's creation narrative—is a flash of light within whose expansion physical bodies eventually coalesce. This, however, is a broad similarity. Of greater import is the idea that whatever happens first, in an originary sense, defines what is possible thereafter. Thinking in this vein, J. T. Fraser traces the constancy of the speed of light back to the big bang, insisting that this first velocity "has retained a unique and invariant relation to all states of motion that have subsequently become possible."[11] Nothing can exceed the speed of light because that primitive flash of light set bounds on all future states of motion, and it did this by being the defining moment of creation. Nothing existed before or beyond it; it did not occur in a preexisting spacetime field. On the contrary, that field and all the material bodies that now populate it emerged in the aftermath of first (originary) light. It makes sense, then, that *they* take their cues from light, from its motion, rather than the reverse.

Put differently, although the big bang occurred long ago, it is not completely buried in the past. In virtue of its capacity to originate the spacetime regime while conserving its primal, pre-regime (spaceless, timeless) integrity, the big bang shows up in the conditions light imposes on the dynamical properties of material bodies. Not only that, it shows up in light itself, which does not wholly conform itself to familiar modes of action presupposing space and time.

As Fraser puts it, "All instants in the life of the photon are simultaneous."[12] Here, of course, he is thinking of time dilation, which for light is complete and therefore fully subversive of time (and, by implication, space). Seen this way, light is a throwback to the first moment of light, which is the moment of creation; its indifference to space and time is aboriginal, and, it would seem, timelessly operative. After describing how photons circumvent space and time in physical experiments, John Wheeler proposes that each photon constitutes "an elementary act of creation" as it strikes the human eye or some other instrument of detection. He then asks: "For a process of creation that can and does operate anywhere, that reveals itself and yet hides itself, what could one have dreamed up out of pure imagination more magic—and fitting—than this?"[13] Yet, as just noted, the resources for this sort of action are intrinsic to light because, to follow Hermann Bondi, light "cannot change once it has been produced, owing to the fact that it does not age, and therefore it must remain the same."[14] Granted, the material cosmos has aged or evolved since the big bang, but light, its primal stuff, has not. There is a sense, then, in which light, having once contained the cosmos as a dimensionless point, continues to hold it intact.

To fully appreciate this aspect of light—the way it absents itself as a thing per se from the spacetime regime while functioning as a principle of cosmic unity or creation in that regime—we turn to Thomas Torrance. Keying off both modern physics and Christian theology, Torrance has seized upon relationality as light's essential nature: it shows up only in relation to or conjunction with other things. One of these other things is mind, which for Torrance is harmonious with light. As Torrance remarks, because of light there is "a structural kinship between human knowing and what is known."[15] This prompts the suggestion that light's relationality keeps it from being lifted out of the seeing, knowing experience. Thus it remains intimately associated with, even implicit in, all we normally deem it to be apart from.

Light's Relationality

Torrance builds his "theology of light" around Christian understandings, but one need not be Christian to appreciate his insights. Not entirely his own, those insights draw on humankind's age-old fascination with light as well as modern physics' more recent revelations and puzzlements. Torrance argues that relativity theory sparks the understanding that although individual objects may be deemed *parts* of a whole, nothing is *apart* from the whole. Even the knowing mind enjoys "a structural kinship" or congeniality with "what is known."[16] This kinship, expressive of all of nature, is constituted by light. We live in "a universe of light," says Torrance, because light is a universal ordering principle.[17] Nothing can exceed its immense velocity, and, more cogently, the speed of light regulates the behavior of all material bodies in an absolutely impartial way. Aside from being an earnest or witness of God's unconditional love and faithfulness toward his creation, this impartiality gives us a unified world, a cosmos, by bringing all within the embrace of a single law or principle that never suffers contravention. In science, this law finds expression as the constancy of the speed of light.

Torrance insists that light's constancy, though inexplicable from the standpoint of common sense and everyday experience, signifies or reenacts God's relationship vis-à-vis his creation. According to the Bible, God is "no respecter of persons";[18] he "maketh his sun to rise on the evil and on the good, and sendeth rain on the just and on the unjust."[19] In other words, God's unconditional love preempts any suggestion of cosmic favoritism or privilege. Similarly, the invariance of the speed of light preempts the possibility of privileged observation in the cosmos. While this latter fact entails the relativity of all reference frames, it yet secures the absolute inviolability of a universal constant against which those frames adjust themselves. That is, the speed of light, by reason of its independence from the motion of specific material bodies, brings all such bodies into a single coherent scheme. As Torrance, citing Louis de Broglie, proposes, light fulfills its obvious illuminative function by disclosing a vast interconnectedness among things: light shows up in the wide coherence or intelligibility of the optical experience.[20] This disclosure, however, is inseparable from the fact that the dynamical (spacetime) properties of all material bodies conform themselves to the speed of light, so that all appear on the same visual page, as it were.

To give the same point a slightly different emphasis, light's behavior is unaffected by the varying motions of material bodies, and this universal steadfastness in the face of local change brings all those motions to a common existential plane: local effects achieve universal coordination. Consider the converse of the postulate that the speed of light is constant irrespective of the motion of its source or the motion of observers: from the point of view of a light ray, all material bodies move at the same speed—the speed of light minus the speed of light. That is, once the invariance of light's velocity is directed back at light-illuminated bodies, nothing moves faster than anything else: everything is captured or encompassed by light in an absolutely equable way. For Torrance, this equality suggests God's grace: to be illuminated or light-awakened is to share equally in God's superabundant love.

One is reminded here of Galileo's comment that "God and Nature are so employed in the governing of human affairs that they could not apply themselves more thereto if they truly had no other care than only that of mankind." To secure this thought, Galileo notes the action of light: "And this, I think, I am able to make out by a most pertinent and most noble example, taken from the operation of the Sun's light, which . . . in ripening that bunch of grapes, nay, that single grape, . . . does apply itself so that it could not be more intense, if the sum of all its business had been the maturation of that one grape."[21] Because of light, it is as if God and nature focus their entire attention on each single detail of the world: that is both the everyday lesson of light, noted by Galileo, and an idea growing out of the foundations of modern physics, according to Torrance.

For Torrance, "physical" or "created" light is an introduction to the "uncreated" light of God. Einstein's dismissal of the luminiferous ether turned physical light into an auto-referential reality— something that, like God, is its own "ultimate ground."[22] Evidently, light is free of physical matter in two respects, each of which is a variation of the other. First, light is indifferent or impartial to the motion of material bodies; second, light as "its own thing" does not depend on a material medium for its propagation.[23] Like God, it preserves its universal nature by acting in a widely relational way that preempts local or individualistic distinctions. Humans, of course, can focus on particular things, but they do so by fragmenting the optical expanse and turning shards thereof into localized, self-standing objects. Oft forgotten in this process, however, is the unitary visual experience and the relational character of light.[24]

Light relates to other things by interrelating them; it follows that in light's economy nothing may be free- or self-standing: all are mutually implicated. A similar rationale might be said to inform God's impartiality toward his creatures: the human tendency to single out or isolate is precluded by a clear vision of interrelations. And because of the cosmic distances involved, that vision is vast. But it is also compactly integrated by light's capacity to effect what John Wheeler calls "zero-interval linkages between events near and far."[25]

That humans cannot see the interrelations, or easily infer their existence, is not surprising, for both God and light are invisible. Torrance notes that from St. Augustine onward, Christian theologians have remarked on the uncanny fact that "we cannot see light itself but only things lit up by light."[26] Or, expressing this insight even more paradoxically, "in a strange way physical light is at once a darkness in itself and yet the source of brightness all round it."[27] God, says Torrance, is similarly operative in the world— "unapproachable and invisible, but . . . [as] illuminatingly present in the world of thought as the created light of the sun is in the world of sense."[28] These claims, while provoking rich theological comment from Torrance, are less explicitly tied to physics. He merely observes that while light travels much too swiftly for us to see, the constancy of its velocity keeps it from being overtaken by human observation: a fixed interval separates us from it, no matter our speed. And since we cannot slow light down with our own speed, it forever bids us onward into change, repentance, and new understandings of God.

One may refuse to go the full distance with Torrance's outlook. That said, it, in a manner consistent with modern physics, fosters a heightened awareness of relations. And this in turn prompts the thought that if a principle of cosmic relationality or unity exists, we should not expect to discover it as a *part* of the cosmic whole. It would always reassert itself in a relational way. Light seems to fit this description. Keying off Einstein, Torrance proposes that light fulfills its biblical function as a principle of creation by playing an elemental and universal role in harmonizing events in the cosmos. By going all the way down, by never stopping to announce itself as something apart from all else, light's relationality interrelates or unifies the whole.

Questions remain, of course. One may ask, for example, the following: if light supposedly plays a unique role in the cosmos, why then do electrons and other material particles *also* exhibit some of the strange behavior typically associated with light? There are several ways to respond to this question. Let me remark here that if light's uniqueness consists in its unparalleled relationality, then the very attribute that sets it apart from other things also brings it into relation with those things. Torrance, reflecting on Einstein's unification of classical mechanics and Maxwell's electromagnetic theory, calls light "the most refined form of matter."[29] This does not simply mean that light is the upper end of a matter-light continuum; more fundamentally, it refers to light's uncanny ability to integrate reality by means of relation. Distinctions that would otherwise hold absolutely are thereby brought into conjunction and blurred, whether they be mental dichotomies such as wave versus particle, different physical (spacetime) locations, or, light's own uniqueness vis-à-vis the material world.

6

INTERNAL RELATIONS

Into the whole how all things blend
Each in the other working, living!

>   Johann Wolfgang von Goethe,
>   *Faust, Parts One and Two*

The last chapter depicted light as a relational phenomenon. I have suggested that light is not a thing among other things; it is a unifying relationality that enables the identification of one thing vis-à-vis another. The constancy or steadfastness of light is a fixed basis from which various things emerge. Absent this basis, there would be no platform for the origination of *relative* difference, which is difference grounded in underlying sameness or constancy. There would be no cosmos, no realm in which various things show up on the same page, so to speak, and collectively hang together.

But to say that things "hang together" in the cosmos is to put the matter too weakly. In light's economy, things don't just hang together: they achieve mutual immanence. Something like Indra's net of pearls occurs, although this metaphor, as noted, fails to capture the full reality of relational light. In this chapter we try to go further by reviewing David Bohm's implicate order and Karl Pribram's holographic brain. These outlooks, both of which posit the mutual immanence of supposedly separate entities, prepare us for the revolutionary concept of internal relations.

Bohm and Pribram

As children most of us experienced the surprise that comes with unfolding a folded piece of paper from which a small section has been cut. As the paper unfolds or expands, what we thought was a single cut becomes a pattern of several cuts. This is one of several ways Bohm illustrates his implicate order. In that order things fold into each other in some aboriginal, unitary fashion. They then emerge into the explicate order, the world of everyday experience, to become separate objects—at least, apparently so. Knowing only the explicate order, most people take it to be the sum total of reality. Puzzling over issues in physics, however, Bohm felt obliged to posit the implicate order.

One of those issues was quantum entanglement or nonlocality. It is easy for us to see that, in some sense, two or even ten cuts in an unfolded sheet of paper are the same cut. That is because the paper is distinct from us (in the explicate order, at least) and we therefore can go through the entire process of folding it, cutting it, and then, as we unfold it, witnessing the one-many transformation. But at a deeper ontological level all this would not be obvious. Subtly textured into the explicate order, we would have difficulty getting intellectual leverage on it. And if we were to witness a comparable sort of sameness in the face of difference, or co-locality in the face of apartness, it would be puzzling to us. Most people would not surmise that quantum entanglement might arise from the explication or unfolding of a prior unity or all-in-all. That is, they would resist the thought that quantum-entangled particles are the smallest tip of a deep-seated cosmic entanglement or enfoldment.

Some of this resistance stems from our awe of space and time. These two great separating modalities seem utterly real, even omnipotent. But the greater reality, the reality to which space and time are subordinate, is light speed constancy. Knowing this and the overall manner in which modern physics challenges Newtonian thought, Bohm felt no compulsion to accord space and time absolute status. They hold sway in the explicate order, but that realm is derivative of the spaceless, timeless implicate order. And into this latter order, Bohm insisted, collapses the once venerable scientific project of grasping the world as a mechanistic system composed of separately interacting parts: "Ultimately, the entire universe (with all its 'particles,' including those constituting human beings, their laboratories, observing instruments, etc.) has to be understood as a single undivided whole, in which analysis into separately and independently existent parts has no fundamental status."[1]

Another of Bohm's illustrations is a carpet pattern: "In so far as what is relevant is the pattern, it has no meaning to say that different parts of such a pattern (e.g., various flowers and trees that are to be seen in the carpet) are separate objects in the interaction." "Similarly," he continues, "in the quantum context, one can regard terms like 'observed object,' 'observing instrument,' 'link electron,' 'experimental results,' etc., as aspects of a single overall 'pattern' that are in effect abstracted or 'pointed out' by our mode of description."[2] To apprehend a pattern is to apprehend at a glance a *tout ensemble* or all-in-all, and the pattern is lost once one defines its parts as separate entities rather than mutually informing elements of a larger system. According to Bohm, parts of a pattern don't even interact; the thought of interaction arises as we *explicate* the parts, lift or abstract them from a seamless unity. Then those parts *must* interact if the cosmos is to be something more than a cacophony of random (fully unrelated) events. But with this shift in thought, which sees the universe as a mechanistic system, it becomes hard to re-see the larger, deeper pattern of things wherein various entities cohere and interpenetrate. Like those trying to reconstitute Humpty Dumpty, we have difficulty reintegrating the primal whole.

For a culture steeped in the Newtonian metaphor of a clockwork universe, nothing is more natural than analysis into parts. At the same time, however, there is something profoundly artificial about such analysis. Think of the Big Dipper, which most people can readily point out in the night sky. Here are seven stars that the ancient Greeks lifted out of the sky as a separate part or constellation, and they did this by playing dot-to-dot with seven points of light to create a particular pattern. To see that pattern is to catch at one glance a celestial dipper. But of course there is something arbitrary about this. Other cultures constellated the stars differently, and so we can say that the Big Dipper is a human construct, not a naturally occurring feature of the world. Continuing in this Newtonian line of analysis, we might argue further that the seven stars that form the Big Dipper are, at bottom, separate entities. Of course they are gravitationally interconnected, but gravitational force is not intrinsic to their being. It is something extra; it is an outside force or relation.

Bohm would probably agree with a lot of this, but he would add that our propensity to organize natural phenomena into patterns should not be dismissed so quickly. Patterns bring meaning into our lives, and science, even as it dismisses old patterns, reinvests nature with new patterns. Not only that, but the patterns we find in the world echo the implicate order, just as the pattern of cuts in an unfolded sheet of paper echoes an originary cut. For Bohm, then, relational patterns are as important as the things that play into the composition of patterns. And one of those things is the human mind—it does not stand on the sidelines as a separate entity or magisterial spectator. It is already patterned into nature before it begins to discover nature's patterns.

Yet another illustration of the implicate order involves laser holography (see figure 6.1). This illustration is particularly striking because the effect it produces relies on and demonstrates the holistic action of light. A holographic recording is taken by directing coherent light from a laser toward a beam splitter. Half of the light goes through the beam splitter and hits a photographic plate. The other half is reflected onto an object of photographic interest and then bounces onto the plate. Meeting at the plate, the two streams of light produce an interference pattern. To see the resulting image—one stands behind the plate while a single beam of laser light illuminates the pattern from the other side. A three-dimensional image—a hologram—appears. This in itself is surprising, but other surprises quickly emerge. As the viewer steps around the image—that is, changes her angle of vision—the image changes as if she were moving around the object itself. Moreover, should the plate

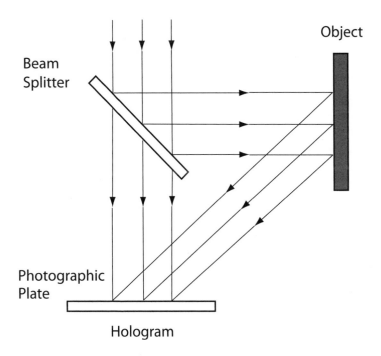

Figure 6.1. Laser light is split in two. Half passes through the beam splitter and half is reflected onto an object and then onto a photographic plate. As the two halves interact at the plate, an interference pattern develops that then can be viewed as a hologram.

be cut in half, she will not see half the image (as she would with a photograph) but the entire image, albeit somewhat less distinctly. Every part of the interference pattern encapsulates or encodes the entire image.

Bohm cites holography as evidence against the familiar model of reality that assigns fundamental value to parts. As noted, one can take a conventional photograph and cut it—and its image—into parts or pieces, and this leads us to believe that reality itself can be so cut up. But even though a holographic plate can be cut into parts, the image cannot. Granted, it loses visual definition with each cut, but the image as a whole remains intact and thereby implies indivisibility in the face of seeming divisibility. If photography with its flat images points us toward parts, holography with its rounded, more true-to-life images points us toward wholes.

While these illustrations are meant to be analogies of the implicate order, they do not, of course, prove its existence. Still, they suggest that parts do not tell the full story of reality; wholes also figure in, and probably at an originary level. This, at any rate, is Bohm's thinking, and he argues that quantum entanglement is an instance of human curiosity touching down at that level. We perform experiments involving local particles, particles we take to be obedient citizens in the spacetime regime, only to discover their nonconformist, nonlocal character. In brief, under certain circumstances, assumed parts, like opposite ends of a carpet pattern, become "unbounded portions of the whole" rather than self-bounded parts.[3] When this happens, we cannot pin those parts down to precise locations: they get holistically diffused or distributed throughout the entire system.

Bohm's implicate order played into the thinking of Karl Pribram, a neuroscientist who posited a holographic brain after he and other researchers failed to localize the brain's memory function. The early-twentieth-century model of the brain assumed that different regions correspond to different brain functions. But research showed that removal of or injury to the memory region—so assumed—did not wipe out memories. It was as if memories were dispersed throughout the brain. Similarly, other brain functions could not always be localized. Of course, this in itself did not demonstrate that the brain behaves holographically, but Pribram felt that holography helped make sense of his research data, and he expanded his understanding of quantum theory by exchanging ideas with Bohm.

Among other things, Pribram found that vision, hearing, and bodily movement can be mapped with holography onto a common mathematical basis known as Fourier transforms. These enable the conversion of patterns into waves and vice versa. Pribram proposed that the brain automatically converts incoming, mutually interfering wave fronts into the patterns or images we see in the world. Thus we sense apparently stable (particle-like), self-isolated things like chairs and rocks, never suspecting that such have emerged from a more fluid, more far-flung, and more elementary wave reality. This wave reality coincides with, or is analogous to, Bohm's implicate order. It is also akin to the wave interference fringes on the holographic film plate that subsequently get converted into a distinct image of an object in this world, the explicate order.

My intent in rehearsing Bohm and Pribram's hypotheses is to suggest that even if these thinkers may be wrong in part, they are still right in breaking with the mechanistic model of reality that elevates parts at the expense of wholes. Further, given this elevation or exaltation of parts, it becomes very difficult to understand light and light-related experiments that open out onto nonlocality. If nothing more, Bohm and Pribram loosen up our thinking in a good way. We are now ready to think about internal relations.

Internal versus External Relations

Perhaps the best way to understand internal relations in general is to interrogate our unquestioned commitment to the idea of external relations. No one denies that various things are interrelated. The question at issue, however, is whether relations are (1) merely incidental to those things or (2) constitutive of their being. The general response is (1): things exist on their own and then get related to each other. A particular rock, for example, exists independently of all other things (e.g., other rocks, trees, etc.); it does not borrow its existence, via relationality, from those things.

This is the concept of external relations, and it is the lens through which we typically see and understand the world—a world that then makes sense as a mechanistic system. Think of two billiard balls in collision. Because the balls are deemed self-existing entities, any collision—any relation—between the two must be unessential or external to their fundamental nature. Accordingly, the balls bounce off each other and proceed unchanged on their way. Granted, their momentum is altered, but that alteration took effect at the moment of collision and then, according to Newton's first law of motion, simply persists in virtue of the lifelessness or inertness of the affected bodies. That is, the balls are not enriched or enlarged ontologically on account of the collision. They remain blank within, and all interaction or relation between them is incidental to that irremediable blankness. Simply put, they have no inner state or subjectivity upon which relations can impinge.

This is the outlook that informs modern thought, and for many it seems unassailable. Why, after all, would anyone want to ascribe inner subjectivity to a billiard ball? But no metaphysical position of this type is beyond criticism. Consider that some things clearly borrow their existence, by way of relationality, from other things. We may say that summer is warmer than winter, but it is just this temperature relation (among other things) that enables the reality of summer. Winter and summer are different seasons that nevertheless lean into each other, and both gain in richness or intensity as the contrast between the two widens. Here, then, is an example of internal relations: the relation is internal to or constitutive of the thing being scrutinized.

If we widen the relational web of summer sufficiently, we include ourselves, and then the idea of internal relations comes home. To some significant extent, each person is shaped by the summers she has experienced: those summers did not leave her unchanged. Further, each summer, by being implicative of other seasons and therefore inclusive of them, in turn triggers wider associations. Where this expanding cascade of relations ends no one can say; it is coincidental with life itself. No person is a freestanding entity; each is a nexus or knot in the vast relational web.

In discriminating between internal and external relations, much depends on level of detail, the focal adjustment of one's analysis. At a higher or more macroscopic level external relations, or the assumption thereof, may coincide perfectly with what one expects to observe: mindless, self-contained entities in mechanical interaction. But details begin to wash out with greater resolution or tighter focal adjustment, and this strips the entities of interest of their well-detailed self-containment. Or, to follow the prevailing interpretation of quantum physics, at this more fundamental level entities of interest (electrons, etc.) intrinsically lack the kind of detail associated with large-scale, apparently self-contained objects like billiard balls. And without that detail, nothing exists to neatly circumscribe their reality in space and time.

If this sounds counterintuitive, it really is not. Bohm offers the analogy of language. One can analyze a story in the hope of understanding it better, but at some point—at about the level of individual words and letters—the analysis process backfires. Understanding or meaning begins to fuzz out. This, he says, is "because the 'intrinsic' nature of each element is not a property existing separately from and independently of other elements, but is, instead, a property that arises partially from its relation with other elements."[4] Here is where internal relations with their inter-mutuality of elements prevail, and so no element can be captured in isolation. Each intrinsically references or embodies other elements.

It is easy to see that each attitude, whether external or internal relations, fosters a different view of physical reality. The former works from the assumption of inner blankness, the latter from inner receptivity. Of course the concept of external relations is not without utility, but from here I follow Alfred North Whitehead, a mathematician and philosopher who argued that while that concept promotes the agenda of mechanistic science, internal relations are the living threads of everyday experience. We are not left unchanged by our relation-based encounters with other things.

Whitehead called his outlook a "philosophy of organism" because it transposed the life-experience of internal relations to a cosmic basis. Those relations, being subjective, always entail *feelings*—perceptions, moods, dispositions, and, at more fundamental levels, unconscious apprehensions and processes that enable conscious experience. Without these feelings, we are back to the realm of external relations: lifeless, self-contained bodies unfeelingly crashing into each other, the sum of which could never account for our primitive experience as context-inclusive beings. From the very start, our experience is relationally configured and thereby outwardly textured into a "world" or "cosmos." This realization, said Whitehead, "springs from direct inspection of the nature of things as disclosed in our own immediate experience." Therefore, he continued, we "know that in being ourselves we are more than ourselves"; we "know that our experience, dim and fragmentary as it is, yet sounds the utmost depths of reality."[5]

The point I wish to make is that Whitehead deliberately grounded his philosophy to what we know best and most intimately—life—rather than to what we merely choose to infer from observation—lifeless mechanism. Feeling himself a living being, he aimed for an outlook that brought him forward as such, a being feelingly related to myriad other things. And as he developed the outlook and turned it into a cosmology, he never lost sight of the living experience that nourishes theory construction and the way that this experience spills single, seemingly self-contained individuals into a cosmos shared by other people, rocks, clouds, stars, and so on.

Whitehead is famous for his remark that "[t]he safest general characterization of the European philosophical tradition is that it consists of a series of footnotes to Plato."[6] What he said thereafter is equally interesting:

> [I]f we had to render Plato's general point of view with the least changes made necessary by the intervening two thousand years of human experience in social organization, in science, and in religion, we should have to set about the construction of a philosophy of organism. In such a philosophy the actualities constituting the process of the world are conceived as exemplifying the ingression (or "participation") of other things which constitute the potentialities of definiteness for any actual existence.[7]

Here ingression or participation refers to our inherent or internal relation to other things. To a greater or lesser degree, all things figure into our being and we into theirs. Hence the universe is "a solidarity" of interconnections.[8] Of course mechanistic science also speaks of interconnections and mutual attractions, but these are impersonal and consequently cater to the premise of grasping nature in a fully objective, unfeeling manner. Whitehead saw the flaw in this premise and sought to create an outlook that coincides with the way we experience the world before unfeeling analysis, or the pretense thereof, kicks in. A great deal of confusion, he proposed, could be cleared up if only we could learn to undo such analysis.

As a case in point, he noted that "[s]cience and philosophy have been apt to entangle themselves in a simple-minded theory that an object is at one place at any definite time, and is in no sense anywhere else."[9] While this may be the attitude of science and even common sense as it has been shaped by science, it "is not the attitude of language which is naively expressing the facts of experience. Every other sentence in a work of literature which is endeavouring truly to interpret the facts of experience expresses differences in the surrounding events due to the presence of some object."[10] What this implies, concluded Whitehead, is that "[a]n object is ingredient throughout its neighborhood, and its neighborhood is indefinite."[11]

Put this way, the idea sounds at once underwhelming and astonishing. Of course the flowers in my front yard are not *just* there if they are also affording aesthetic pleasure to a neighbor across the street. But if we concede this much—that the flowers' presence is coincidental with their impact on other things around them—where, then, do the flowers cease to be? Conceivably they do not; to some degree, however minute, they are participatory in all else. The difference they make triggers other differences ad infinitum, not mechanically or externally but by becoming ingredient in those differences so as to secure the flowers' ongoing, ever-culminating presence.

This may sound poetic, and to some extent it is, but Whitehead believed that the testimony of poetry needed to be heard, particularly in view of the "strained and paradoxical . . . view of nature which modern science imposes on our thoughts."[12] That view disparaged the witness of everyday experience by reducing nature's components to lifeless blanks solely given over to external relation and mechanical interaction. Nature thus becomes "a dull affair . . . merely the hurrying of material, endlessly, meaninglessly."[13] But if nature is a dull affair, why isn't science also? Does it seem odd that nature, dull as it is, would at some point begin to take a scientific interest in itself? How do curiosity and fascination bootstrap up out of elemental monotony?

In blinking away a great deal of ordinary experience, science had worked itself into a predicament regarding location. This stemmed from what Whitehead called the fallacy of simple location, the erroneous view, noted above, that things exist at specific places at specific times and nowhere else. To go back to the flowers in my front yard: if they exist only there and only during the summer, how do they register themselves in my memory now that summer has passed and I am several miles away from home? Put more generally, if the story of every event fully unfolds at the simple location delineated by science, how is it that we can talk about events when we are spatially and temporally distant from them? That we do talk about them, that they do provoke our interest and thus alter our being, gives the lie to external relations. It also gives the lie to the premise of simple location. If that were true, nothing about an event would inherently refer "to either the past or the future" and "memory . . . would fail to find any justification within nature itself."[14]

Something about events, in brief, must inherently tie them into spacetime regions beyond the local coordinates assigned them by science; otherwise they could not be picked up in those regions by other entities. The tie-in is internal relations, relations whose nature is not distinct from the events tied together. Thus, at a level beneath the radar of mechanistic science, events themselves intermingle, irrespective of space and time intervals. To follow Whitehead: "In a certain sense, everything is everywhere at all times. For every location involves an aspect of itself in every other location. Thus every spatiotemporal standpoint mirrors the world."[15]

This is not a mystic speaking. Whitehead felt he had arrived at this conclusion from rational principles and from a consideration of everyday, pre-scientific experience. He wrote: "If you try to imagine this doctrine [the mutual immanence of all things] in terms of our conventional views of space and time, which presuppose simple location, it is a great paradox. But if you think of it in terms of our naïve experience, it is a mere transcript of the obvious facts."[16] Although physically confined to a single location—one might say a simple location—each person is, in a fully involuntary way, perceptually projected into a vast milieu that fades away "into the general knowledge that there are things beyond."[17] That projection puts us in meaningful touch with other things, both seen and unseen. What is more, it is there from the start; it is human experience, human being, before materialistic science scaled it back to simple location.

Summing Up

Three final points need to be made. First, in response to Whitehead's claim that "every spatiotemporal standpoint mirrors the world,"[18] we should remind ourselves that light speed constancy uniquely centers each person at the middle of the world even though each occupies a different spatiotemporal standpoint. If you produce a spark, its light moves away from you in all directions at an absolutely constant rate—the speed of light. But if I see the spark, the same thing happens for me, even though I may be far from you and even though I may be moving away from you at a high rate of speed: light from that spark immediately expands away from me omni-directionally at the speed of light. So while I see the spark as faraway, its light is not behaving in every respect as if I were on the periphery of things. Thus each observer occupies a central point around which the world uniquely pivots. Each "mirrors the world" from a central location.

The second point follows from Whitehead's claim that, despite the fallacy of simple location and the way it now muddles our thinking, people sometimes sense the mutual immanence of all things. William Wordsworth is a case in point. He wrote that he felt "something far more deeply interfused,"

> Whose dwelling is the light of setting suns,
> And the round ocean and the living air,
> And the blue sky, and in the mind of man;
> A motion and a spirit, that impels
> All thinking things, all objects of all thought,
> And rolls through all things.[19]

This ultimately is what Whitehead is trying get at, albeit in a less poetic, more philosophically rigorous manner.

Finally, there is the stumbling block of Whitehead's insistence that nature is organic rather than mechanical. Does this mean that a rock or atom is feelingly responsive to its environment? In a word, yes, although at these levels responsiveness need not imply consciousness or self-creativity (agency). Still, many moderns will find this an incredible proposition. Whitehead, however, argues that the contrary thesis—the clockwork universe composed of lifeless cogs— is not a palpable, self-evident fact of nature. It is, instead, an idea abstracted from an interpretation of nature that prizes mechanism and materialism. Not that this abstraction is wrong in all cases. When limited to its obvious function, that of grasping and exploiting nature's mechanistic aspects, the abstraction has demonstrated its utility and epistemic legitimacy many times over. As Whitehead says, "The narrow efficiency of the scheme was the very cause of its supreme methodological success." Nevertheless, "when we pass beyond the abstraction, either by more subtle employment of our senses, or by the request for meanings and for coherence of thoughts, the scheme breaks down at once."[20]

As indicated, Whitehead's corrective was "direct inspection of the nature of things as disclosed in our own immediate present experience. There is no parting from your own shadow."[21] This shadow, elemental everyday experience, connoted for Whitehead a larger reality than that portrayed by materialistic science. It connoted life, and his philosophy of organism was an attempt to sketch "a mere transcript of the obvious facts" embodied in that connotation. Unlike vitalists, he did not posit a spiritual or vitalistic life force that would then govern lifeless elements; rather he argued that if we attend to what we know firsthand—the stubborn facts of everyday experience—life registers as a ground state from which we develop anti-life propositions, that is, propositions such as external relations and simple location.

Much more could be said about Whitehead's philosophy as it bears on issues in science. For now it remains to point out that his criticism of simple location, delivered early in the twentieth century, portends the later experimental determination that paired particles do not behave according to what Einstein called "the locality principle." Instead, they timelessly interact (if indeed "interact" is quite the right word) across arbitrarily large space intervals. Some thinkers hold that Whitehead, by putting primary emphasis on experience, provides a foundation for solving the puzzle of nonlocality.[22] Others, innocent of Whitehead but compelled by nonlocality to think deeply about relations, have independently concluded that "the relationships [in a system] between the constituent parts are 'internal or immanent' in the parts . . . as opposed to relationships that are external to the parts."[23] Be that as it may, here I am more modestly suggesting that the idea of internal relations can open a path to grasping light's deeply relational action. That action violates the requirement, imposed by external relations, that different things remain apart (confined to their simple locations) while interacting with each other. By definition, internal relations also violate the apartness requirement.

7

LIGHT IN A VACUUM

Yet light, it seems, is always ready
with another surprise.

Ralph Baierlein, *Newton to Einstein:*
The Trail of Light

Before going further, we need to remind ourselves that the speed
of light as it normally figures into physics is the speed of light in a
vacuum. Light slows down as it moves through physical media like
glass and water, and so its speed is not constant under all circumstances. When, however, light travels in *vacuo*, all inertial observers
measure its speed at the same rate, and it is this speed—denoted by
c in science—that is deemed a constant of nature.

Light's relation to a vacuum is emblematic of light's nature,
emblematic of something that flies beneath the radar of experience
until striking *other* things. Then it flashes into experience, and,
it seems, into existence. Empirically speaking, that flash is fully
coincidental with light's reality. We get no hint of light's existence
upon looking into a vacuum, even after shining light therein. But

add a material object to the vacuum and light immediately informs us of its own existence by visually announcing the existence of the object. To all appearances, light coexists as comfortably with the nothingness of the vacuum as it does with the somethingness of material entities. Consistent with its expansive nature and what we have learned about light in other contexts, light straddles another divide: in this case, the divide between somethingness and nothingness.

Modern science has learned to shy away from sharp demarcations. Once waves and particles were starkly dichotomized, while space, time, matter, and energy all fell into different categories. Now these boundaries have been erased, and the Cartesian mind-matter divide is under assault. Not only that, but the once clean demarcation between being and non-being no longer commands assent. This is a point to which we shall return. First, however, let us assume that being is distinct from non-being and see where that lands us. Let us ponder the relation of something—light—to nothing—the classical (pre-quantum) vacuum of nineteenth-century physics. I propose that even before modern physics began to blur this divide there was, at least from an empirical standpoint, something quietly puzzling about Einstein's dictum that the speed of light in a vacuum is constant in all reference frames. What would it mean to witness light moving in a vacuum? Trying to answer this question is another pathway into the realization that light is infinitely more than we know it to be.

A Basic Incongruity

For Einstein, the vacuum entailed the absence of the luminiferous ether or any other material entity. Nothing like the ether could support the propagation of light, for that would turn the speed of light over to the properties of a material medium, and this, in turn, would imply different relative speeds for different observers. Light would once again be ensnared by physical matter.

But what does it mean for light to be completely free of physical matter, to exist in a vacuum? The light we see about us—reflected, scattered, refracted light—has been stopped, altered, and slowed by material bodies and media. Like many scientific laws, light speed constancy seems an idealization not easily achieved amid everyday circumstances: no one has ever seen or made a perfect vacuum. Near-perfect vacuums do, however, exist in outer space, and so confirmation of Einstein's second postulate is not difficult. An astronaut need only send a light beam toward a mirror and wait for its reflection. If she knows the distance to the mirror and can measure how long the beam takes to make its round trip, she can calculate its speed. And by varying her inertial motion from one experiment to the next, she can further establish that her own motion does not affect the speed of light—that speed is constant across all measurements. This is exactly what Einstein proposed.

All this is straightforward, but let us note that the astronaut never actually sees light *moving* in a vacuum. She works from known facts to calculate its rate of motion, but she does not see it moving in intermediate space—the vacuum itself—as we might see a baseball moving through space. This limitation is not her fault or the fault of the experiment. It is inherent in the nature of light. As thinkers since Plato have insisted, light per se is not available for objective viewing—it shows up only while interacting with physical bodies.[1] And because the vacuum is empty of objects, light does not announce itself therein. Any claim regarding its motion in a vacuum, therefore, can be nothing but inference. Very good inference, perhaps, but inference all the same.

This point is critical and one to which we shall return. Of course we see reflected and scattered light, but never do we see light in the absence of material bodies—in a vacuum. One gets a sense of this by noting that a flashlight shone into the night sky does not visually announce itself, just as a movie projector beam is not seen above one's head in the middle of a theater. Illuminated raindrops or dust particles may be seen, but that is light in conjunction with something material, not light per se. Another example is the sun seen from the moon. It is a material ball of light against the blackness of outer space; it does not visibly radiate light because the moon has no atmosphere to scatter light.

This may seem a minor point, but it prompts the suggestion that in dismissing the ether, Einstein opened a path for a truer, though less comprehensible, appreciation of light. Without a material backdrop, light slips beneath the horizon of visibility, and the idea of light in a vacuum is left without an experiential correlate. Nothing in our experience corresponds to light in a vacuum, for such light cannot flash into experience. Put differently, we see light in virtue of other things, and these things, because they are material, preclude the realization of a vacuum.

So there is something incongruous about the idea of light in a vacuum, at least in so far as experience is concerned. I propose that the incongruity results from light's capacity to bridge mind and world. Or, better said, light is an originary unity out of which the categories of mind and world emerge. As such it precedes those categories and therefore fails to fit cleanly into either one. While most people now locate it solely in the material world (outside the mind), it shows up there never on its own but always in conjunction with other things. Not only that, but it cannot be reduced "to matter or its motions";[2] it cannot be assimilated to the motion of those other things even though it is the agency that brings them to visual experience.

Using physical substances we may block light or alter its speed, and when we do this it shows up in the physical realm. But lacking these substances, we cannot encroach upon it. We cannot alter its speed by moving relative to it; indeed, we cannot even see it moving. We can measure its speed, of course, but only after having interrupted the vacuum by inserting ourselves and our devices therein. And then the measurement is not based on seeing light in transit, but rather on our ability to interrupt its supposed motion so as to calculate how fast it must have been moving prior to measurement.

One may object that an electron, classically conceived as a material particle, can't be directly observed and measured, either. At least one photon must strike it and then rebound back to the observer, having altered one or more of its properties. But in this case we at least see something—an electron that, we surmise, has been simultaneously disturbed and illuminated by a photon. What we don't see in any direct, unequivocal sense is the photon. Granted, according to theory it must impinge upon the eye (or its instrumental extension) to announce the electron, but it is not part of the visual experience. We don't see two things (electron *and* photon); we just see one—the electron—by means of the other—the photon. And when we remove the electron to see the photon in its own empirical right, when we create a light-filled vacuum, we see nothing at all. This, I suggest, is because light, although lending itself to the illumination of physical bodies, is not physical in every respect. Something about it transcends, or perhaps precedes, physicality.

By definition light stands apart from physical matter in at least two ways. It has zero rest mass and it moves at a speed—c—that material bodies, owing to their materiality, can never achieve. But what is more, when moving or existing at that speed, it cannot be seen. It is seen only at lower speeds, and then always in conjunction with the entities responsible for light's fall from c to a lower speed.

Mapped Together

Another way to appreciate the foregoing is to consider what happened to the ether with the advent of special relativity. Although physicists no longer talk about ether, it still makes perfect sense in one way. If light is a wave phenomenon, it should require a substance wherein its wavelike action can occur. The ether was this substance, of course, but it was no ordinary substance. For one thing, since the ether was idealized as a universal rest frame, it had to be subtle enough to penetrate material bodies without being moved or affected by them. For another, the ether had to be incredibly rigid—otherwise it could not propagate light at the tremendously swift rate of c. These two requirements do not jibe easily, for it is hard to imagine an ultra-fine substance a million times more rigid than steel. Even so, scientists pursued the ether optimistically, believing they would eventually detect it. It was, after all, a physical substance.

But, of course, the ether was never detected, and now we are comfortable with that fact. Nevertheless, it lands us in the classical absurdity of talking about light waves without a material medium—an absurdity we are also comfortable with, having been chastened by early twentieth-century physics. Notwithstanding all this comfort, there is, I believe, another way to tell the ether story. Science did not just outgrow the ether. (That did not happen because science did not grow into a more intelligible understanding of nature upon losing the ether.) Rather, owing to light's anomalous nature, the idea of ether collapsed of its own ethereality or ideality. Even more subtle or mind-like than physicists supposed, it etherealized into Einstein's vacuum, where the idea or scientific construct of light moving at *c* resides but nothing else.

This is not to say that ether somehow exists; its material existence is no more real than a vacuum's. The ether's erstwhile materiality, however, supported the pretense that light could be materially differentiated from mind, that light is just another external phenomenon operating in a classical (mechanical) manner already mapped out by the mind. When Einstein snatched away the ether, that classical mode of action could no longer be associated with light, and so light became anomalous because the mapping failed. This much is routinely acknowledged. What I want to propose is that the unthinkability of light waves moving in a vacuum is more than anomalous; it marks the possibility that light is hard to grasp because it is simultaneously mapped into mind and world. It informs or structures our ability to know the world while informing us of the world. This is why the ether had to collapse eventually. Not just because it wasn't there in the first place, but more importantly because it assumed mind-world division, and light is not party to that division. It cannot be cleanly lifted out of things, whether mind or world, because it has something to do with the way things are mapped *together* from the start.

By its very definition, light straddles or precedes the divide between mind and material world. As noted above, light has zero rest mass, so it is often characterized as immaterial. Despite this, it is clearly part of the material world, and that is where science routinely locates it. The difficulty, though, is that it cannot be located in that world in the same way material entities are located. In and of itself it is eventless. That is why light, unlike material bodies, can be put into a vacuum without ruining the vacuum; empirically speaking, it adds nothing to nothingness.

But once something material is added to the vacuum, illumination occurs so as to prompt the inference that light was there all along. Whatever the merits of this inference, the point remains that the idea of light in a vacuum is just that—an idea without an experiential correlate. At this ideational stage, light really is immaterial because its nature coincides perfectly with the absolute immateriality of the vacuum. But when a material object penetrates the vacuum, light penetrates experience by illuminating the object. Idea and experience thus click or conjoin as immaterial light conjoins itself with material reality.

Light is the one thing that can fill a vacuum without violating the vacuum's immaterial integrity. It can be there as Einstein imagined—moving at c—and yet be perfectly indifferent to our scrutiny. We cannot see it moving at c in a vacuum, perhaps because c is the escape velocity for the spacetime regime we normally associate with the material cosmos. In any event, light is certainly part of the material world and, it seems, quietly resident in the immaterial realm of Einstein's vacuum. More than that, however, it seems primordial and thereby discredits the dichotomies we erect to define it. Its elucidation points us back toward an earlier moment—one before the immateriality-materiality or mind-matter distinction took effect.

The Quantum Vacuum

Thanks to quantum theory, the nineteenth-century vacuum, or the concept thereof, no longer exists. No vacuum, that is, shapes up as perfect nothingness. To be sure, it may be void of physical matter, but that need not imply a vacancy of energy—which energy, said Einstein, is equivalent to matter multiplied by the square of the speed of light in a vacuum ($E = mc^2$). Indeed, given light's expansive presence throughout the cosmos, there will, at the very least, be energy in the form of effervescent quantum excitations whose existence is both allowed for and constrained by Heisenberg's uncertainty principle. These excitations are called virtual particles, and it is as if they exist, however briefly and tenuously, between the cracks of the non-chanceful laws of classical physics. They are there in the vacuum, not as observable real entities, so-called, but as probabilities. As such, they pre-exist or sub-exist what may happen, what sort of reality may issue up from a vacuum-filled sea of quantum fluctuations. "All that can appear in reality must be present as possibility—as a state of virtual particles—in a vacuum," writes Henning Genz.[3]

But even before the arrival of virtual particles, physicists were talking about particles existing in superposition states. As Werner Heisenberg, drawing inspiration from Aristotle's notion of *potentia*, put it, such states entail the likelihood that being is linked to non-being by an intermediate reality expressing all possible outcomes.[4] While not every theorist would characterize superposition states this way, most if not all would acknowledge that quantum theory allows for the compounding or mapping together of mind and matter. To follow Freeman Dyson: "The laws [of physics] leave a place for mind in the description of every molecule."[5]

One way of grasping Dyson's point (certainly not the only way) is to reflect that in quantum theory so-called material entities seem to become fully coincidental with their mathematical properties— as if they were nothing more than mathematical ideas. Particle superposition states, for example, are probabilistic (mathematical) descriptions of where a particle may be when measured, but prevailing theory says the probability is ontological rather than epistemic. It is not the case, then, that the particle may be here *or* there in a mutually exclusive way. Rather it is at all possible locations, but more probabilistically in some locations than in others. Similarly, it is moving at all possible speeds, but more likely at some speeds than at others. In other words, our probabilistic understanding or idea of the particle, given to us by mathematics, is perfectly isomorphic with the material particle itself, which, however, can no longer be salvaged as a classical material particle.

Like superposition states, virtual particles are mathematical, and that might predispose us to think of them as unreal, as nothing more than helpful aids in the theory construction process. But this attitude is a bit old-fashioned in that it posits a sharp demarcation between mind and nature. Modern physics discredits this demarcation in many ways, and the quantum vacuum with its sea of virtual particles is one of those ways. As Genz puts it: "It is only in the absolutely empty space of our imagination that no light, no radiation penetrates—that space is as dark as the legendary rooms of Schilda, in the German fairy tale."[6] If something is unreal, it is the perfectly empty vacuum once construed as wedged between nature's materiality and the mind's immateriality. Einstein lifted that wedge just a little when he proposed that the measured velocity of light in a vacuum is constant in all reference frames. At that interface, observer vis-à-vis the unseen agency of observation, materiality and immateriality conflate, for light, never quite reduced to materiality in the first place, is found to have a property of motion that sets it fully apart from material bodies. Something about the world moves with our motion in mind, always staying ahead of us by a constant velocity. Light, I am suggesting, is that something.

8

AMBIENT LIGHT

*Our eyes tell us something about the real nature of
light because they have evolved in response to the
real properties of that entity we call "light."*

John D. Barrow, *Pi in the Sky:
Counting, Thinking, and Being*

Earlier we observed that motion, at first blush one of the most obvious of all experiences, is deceptively complex. This observation
applies with particular cogency to light's motion, which since Einstein cannot be straightforwardly compared to the motion of material bodies. But even before Einstein it was not obvious that light
moves as other things do, or even moves at all. For one thing, no
one had ever seen it moving. For another, light seemed to possess
some property that allowed it to move or change or transmit change
instantaneously. And, to ask a philosophical question, in what sense
is instantaneous movement motion? As we normally use the word,
motion implies the lapse of time, however brief.

Until the seventeenth century people disagreed about whether
light had a finite or infinite velocity. After performing tests that

were inconclusive, Galileo conjectured that light moves infinitely fast: when a cannon fires, we see the flash immediately but must wait to hear the report. Several decades later, Olaus Roemer noted a discrepancy in the appearance of the moons of Jupiter, which he attributed to Jupiter's varying distance from the earth. Working from rough estimates of the distances involved, he offered a calculation of the speed of light that decided the issue in favor of a finite velocity. Subsequent experiments have affirmed and refined this result.

Yet, as we now know, this is not the full story of the speed of light. Thanks to Einstein, light "plays the role, physically, of an infinite velocity,"[1] and so we must again ponder the meaning of light's motion. I propose that its infinite aspect plays into visual experience, but not in the way Galileo had in mind. That is, we do not, in a fully unequivocal way, immediately see distant events—they take time to reach us. If that were not the case, the sequence of events could not flip-flop for different observers. All the same, there is in light an elemental access to the world that militates against our commonplace notions of light moving through space and time.

The Problematic Nature
of Light's Motion

Perhaps the best way to describe this is to say that light's infinite aspect plays solely into pre-theoretical visual experience but hardly at all into theoretical accounts of light's behavior. Open your eyes and you have immediate access to the world; the visual expanse does not enlarge over time. We normally explain this with theory, and the standard theory insists that image-carrying photons—some from the farthest reaches of the universe—constantly impart visual data by their unremitting action on the eye. What is interesting about this account is that it is beyond the reach of confirmation through direct experience. Photons are not seen moving through space and time, and when they strike the eye, they are not seen apart from the images they convey. In theory they are messengers, but curiously the messengers are never seen operating in workman-like ways—carrying or delivering their messages. In theory they give us sight, but do so, it seems, by dropping out of sight.

Consider light in intermediate space, somewhere between a light source and a light receptor. This is not difficult to imagine. In the mind's eye we see light traveling through space as time passes. But this scenario, however reflexive, cannot, even in principle, be empirically realized.[2] Of course we can intercept light en route to its supposed destination, but this is not the same as actually witnessing a traveling photon or light ray. What we observe is the intercept event, not light per se. Freestanding (unreflected, unrefracted, unscattered) light does not show up *on its own* in intermediate space: it is never observed between events (the emission and reception of a light signal, for example), for there is no such thing as an event-free observation. Observations are inserted events that disturb and sometimes truncate previously unobserved processes. In the case of light, observations "terminate" intermediate space by filling it with new events that reveal the existence of light via material bodies—opaque objects, prisms, gas molecules, and so on.

To what may we attribute light's failure to visually announce itself in intermediate space? One option is to trace light's no-show back to epistemic considerations: light is really there, but if it were to announce itself, it would block our view of other things. After all, nothing is closer to the eye than a photon; if we could see photons as distinct entities (apart from the images they convey), what else would we see? In my judgment, this way of thinking about light and vision, while not wholly incorrect, errs on the side of common sense and classical physics. The more daring view—but the one that falls into place with special relativity's counterintuitive outlook—sees light as a fundamentally atemporal, aspatial phenomenon. By this account, light fails to announce itself in intermediate space because, when viewed from within its own economy or reference frame, it is not there. Light's absoluteness, expressed as a constant speed value, signifies some sort of sovereignty from space and time.[3]

It is a truism that metaphorical leaps are taken when talking about elementary particles; we inevitably relate them to familiar objects and thereby risk misconstruing them. As artifacts of the atemporal era—before time and space began to take hold—photons surely challenge our powers of visualization. They are, after all, dimensionless, and their motion is unlike that of material bodies. Despite these sticking points, we often talk about them as if they were little bullets of light, the implication being that they are see-able, self-contained, moving entities. This way of thinking no doubt arose from likening light to material bodies, all of which propagate through space and time.

But on what empirical basis may light be assimilated to moving bodies? Stephen Toulmin writes: "The [ancient] discovery that light travels in straight lines was not . . . the discovery that, where previously nothing had been thought to be, in any ordinary sense, travelling, there turned out on closer inspection to be something travelling—namely, light."[4] That is, looking more closely or directly at light revealed nothing, and the ensuing "discovery" consisted in the realization that one could profitably talk about light in a new way. Toulmin explains:

> Until the discovery, changes in light and shade, as we ordinarily use the words (i.e. illuminated regions which move as the sun moves), remain things primitive, unexplained, to be accepted for what they are. After the discovery, we see them all as the effects of something, which we also speak of in a new sense as "light," travelling from the sun or lamp to the illuminated objects. A crucial part of the step we are examining is, then, simply this: coming to think about shadows and light-patches in a new way, and in consequence coming to ask new questions about them, questions like "Where from?," "Where to?," and "How fast?," which are intelligible only if one thinks of the phenomena in this new way.[5]

Such questions—where from? where to? and how fast?—are now instinctive when thinking about light. But they are also grounded in metaphysical intuition rather than direct apprehension. Certainly there was a gain in intelligibility when light was conceptualized as something that moves through space and time, but at what price? The nineteenth-century concept of light passing through a universal ether was highly intelligible, though mistaken nonetheless. Possibly the notion of light passing through space and time is a kindred fallacy.

Worth noting in this regard are the reservations of two students of the history of Western optics, both of whom questioned the premise of moving light. Vasco Ronchi, a physicist who studied the evolution of optics from antiquity to the early twentieth century, concluded that if we are to learn to talk coherently about light, "we must definitely avoid assuming any distribution of the radiant energy during its *supposed* propagation."[6] For him, both the wave theory and particle theory of light foundered on the mistaken premise of light's motion. Similarly, historian of science Geoffrey Cantor wrote that although light's nature is "beyond our ken," we instinctively find ways to bring it "under the umbrella of matter," although generally without realizing that an imaginative leap has been made. His case in point is "projectile optics,"[7] and our uncritical tendency to embrace literally what can only be metaphor, has, in his opinion, infected our contemporary understanding of light with deep tension.

That tension occasionally surfaces in the expressions of physicists. Nobel laureate P. W. Bridgman directly rejected the idea of moving light, insisting that it is "meaningless or trivial to ascribe physical reality to light in intermediate space and light as a thing travelling must be recognized to be a pure invention."[8] Both epistemic and relativistic considerations motivated this claim. As noted above, we never see light per se "passing through" space;[9] furthermore, in special relativity light by definition travels at the speed of light, and that speed marks a place where conventional understandings of motion—and space and time—break down.[10] How then is light's motion to be conceptualized? John Wheeler offers an answer that splits the difference between conventional descriptions of motion and light's nonconventionality: "Light and influences propagated by light make zero-interval linkages between events near and far."[11] In Minkowski spacetime, time and space values cancel each other out (sum up to zero) at the speed of light, leaving no interval between events otherwise seen (by outside observers) as spatiotemporally separate.[12]

My submission is that the atemporal action of light coincides with visual experience, which would be much less rich were light conventionally temporal. What I wish to suggest, drawing on the work of James J. Gibson, is that while special relativity alerts us in a theoretical way to light's strangeness, that strangeness is also a deeply familiar though routinely overlooked aspect of everyday life. Put differently, special relativity's characterization of light is also descriptive of visual experience.

Gibson's Ambient Light

So far I have given reasons for questioning the premise that light, in a fully unambiguous way, moves through space and time. I have also proposed that light's spaceless, timeless aspect conserves the spaceless, timeless character of the big bang singularity. But living in the aftermath of that singularity and consequently under the sway of space and time, one may well be inclined to dismiss this proposal in the following way: in the far distant atemporal era (before the big bang) there was no space and time for light to move through, but as the universe became more complex, these two great modalities emerged to function as a comprehensive reference frame through which all things, light included, must now move. In other words, what holds at the atemporal era—an era that I have characterized as originary—need not hold at later, more complex eras.

Although eminently commonsensical, this objection plays to Newtonian sensibilities and thereby runs into difficulties. No one contests that the universe, with its spacetime parameters stretching on indefinitely, appears "larger" or more comprehensive than light, but light's rule—the speed of light as a limiting velocity—applies universally, and this may be empirically demonstrated at any time. This fact prompts the suggestion that space, time, and matter (all of which change to preserve the constancy of the speed of light) have yet to ease out of light's originary rule as it now finds expression in the absolute constancy of the speed of light. Over the temporal evolution of the cosmos, light, the foundational principle of cosmic existence, seems to have atemporal staying power: it yet orders the behavior of material bodies in ways that contradict the Newtonian assumption of absolute space and absolute time.

Coming at light from the question of vision, the psychologist James J. Gibson challenged fundamental preconceptions about the roles played by space and time in optical experience. Since Isaac Newton at least, theorists have routinely assumed that visual data move through space and time to reach the eye, whereupon retinal images are produced that then undergo elaboration (interpretation) in the brain. This outlook assigns great value to the brain, because therein the bloom of intelligent vision occurs. The brain takes the small, equivocal, upside-down retinal images and, by means of memory (data storage, enhancement, and organization), transforms them into a well-coordinated mosaic of images. Moreover, even though it seems that visual data would not stream smoothly into the eye—after all, each impinging retinal image corresponds to a discrete moment and is distinct from other images—the brain manages to splice the data together so that no flickering shows up in the seeing experience. This is important, because otherwise the ongoing succession of retinal images from a piano, say, might appear as a rapid succession of many pianos.

For Gibson, this model of vision became more implausible the more he studied it, and the supporting explanations, drawn from a variety of disciplines, did not help. Of central concern was the brain and its remarkable ability to get so much out of so little. Here "little" refers to both the poorly detailed or "impoverished" retinal image and to the more fundamental impoverishment of a world that, by most scientific accounts, is void of spiritual principle. Indeed, the reason the brain has to do so much of the work of seeing is because the standard model of vision reenacts the Cartesian split: the mind is active and experiencing; the material world is inert and non-experiencing. Gibson rejected this dualism as he developed what he called an "ecological optics." Seeing inheres in mind-world continuity or likeness: the like-like linkage between brain and world opens out onto the seamless, expansive visual experience. And with brain-world congeniality or resonance replacing brain-world dissimilarity, there is no need for the brain to cleverly enhance and organize disjointed, poorly detailed visual data. In sum, ecological affinity between brain and world obviates any need on the brain's part to work up a splendid visual presentation from poor sense data: the presentation is not worked up but is a natural consequence of brain-world continuity.

This is not to deny that there is also discontinuity between brain and world—between living, experiencing beings and their natural environment. We stand out against our environment because of our ability to respond selectively to it. We see it as Other and can manipulate it to our advantage, but such manipulation assumes a substratum of similarity from which difference or discontinuity may then emerge. We are, therefore, deeply and primally implicated in the world, not just skimming on its surface in a semi-detached way.

Simple detachment or discontinuity is the great Cartesian half-truth that Gibson sought to remedy. He did so by arguing that perceptual experience cannot be exclusively understood at the atomic or molecular level, although that level might be fine for grasping inorganic substances. We don't perceive atoms or molecules; we perceive trees, buildings, rocks, insects—things more suited to our size. What is more, these things impinge on us, and we on them, in ways that define everyday behavior. I move freely through the transparent air, less freely through semi-transparent water, and generally not at all through opaque surfaces or solids. I pick up and handle tools, smell and ingest food, listen to music, sit in chairs, and shy away from snarling dogs. Of course all these experiences may be broken down into smaller parts (perhaps all the way to the subatomic level) in an effort to explain them, but that is not how they come across. Music, as I listen to it, does not dissolve into separate tones made up of vibrating air molecules. The experience itself is primal, and only in its aftermath—once the mood or feeling has passed—might I be inclined to break it apart. For this reason—because experience per se is irreducible—Gibson let everyday experiences replace unseen particles and processes as the basic building blocks of his theory of perception.

When one adopts this stance, one realizes that perceptual experience is immediately informative of the world. There is no such thing as "informationless perception," wrote Gibson.[13] This thought followed from his decision to take everyday experience seriously. If we choose to privilege atoms, molecules, and the like, we are then obliged to explain how information about the world arises from our interaction with these unseen entities. And the story we tell will involve a similarly unexperienced process of data transmission, assimilation, and enhancement. If, on the other hand, we let experience itself command our thinking *about* experience, we gain firsthand information—the direct witness of experience—that stands on its own merits. In order to do this, however, we must first learn to trust that witness rather than reflexively deliver it over to unexperienced particles and processes.

This is not to say that conceptual abstractions play no role in Gibson's thinking, only that Gibson felt that a theory of perceptual experience should begin with experience and not let primal, pre-analytic reality get fully covered over by abstraction. Approaching things in this way, he hit upon the obvious but routinely overlooked fact of ambient light—light that surrounds and envelops. Physical science sees light as radiant—as radiating from a source—and builds its analysis on that characterization. But, for the most part, the light of experience is ambient, said Gibson. I only occasionally look at light bulbs and generally avoid looking directly at the sun, the great light source in this part of the galaxy. Instead my daily experience involves seeing illuminated objects and the diffused or scattered light of the atmosphere. About me is an interlocking array of surfaces held together by light, the ever-shifting yet structured whole of which constitutes information about the world.

The temptation, again, may be to say that information is had from this array—what Gibson called the optic array—in a second-hand way as images from the various objects converge on the retinas to be processed by the brain. But this garbs the experience with unseen theoretical entities. It also introduces a time element that Gibson felt is not part of the experience. (More on this shortly.) To fully appreciate what is going on with ambient light, one should compare seeing to hearing. As we experience it, sound comes almost exclusively from sources. I hear a voice, a bell, an insect buzzing about, and that is all I hear—not a reverberating ambience or panorama of sound "lighting up" the world. In some situations, of course, sound does reflect off things and I hear an echo, but unless I am blind or blindfolded, that echo does not give me additional information about the world. Blind persons, to be sure, are more attuned to echoes than sighted people, and so for them there is a kind of ambient sound that alerts them to objects and obstacles. Even so, that ambience is much less informative than ambient light. As Gibson puts it, in the case of sound "there is no parallel to the uncountable facets of an array of ambient light."[14]

In a word, while originating from particular things in the natural environment, sound does not obviously "illuminate" the environment—does not put it in a state of circumambient aural reverberation. Neither, of course, does smell, the stimulus for the other distance (non-contact) sense. Odors, like sounds, announce themselves without simultaneously announcing or revealing the milieu in which they exist. Light, however, is remarkable for doing just that, and for doing it so superbly that light's revelatory "opening up" of the milieu was, in many ways, for Gibson light's primitive expression. Put differently, light is coincidental with its expansionistic ambience.

The difficulty for Gibson was that scientific theory routinely treated light as if it were distinct from its revelatory essence, its capacity to "ambiate" the world. Thus, light was portrayed as behaving in a pre-revelatory fashion. It was, for instance, said to radiate from a source and move through space and time before striking the eyes. Assuming that it does do this, however, we do not see it coming and wait for its arrival. More importantly, this modeling of light cannot but oversimplify light's behavior in the everyday world of ordinary experience. Looking at a speck of dirt on the floor, I can't assume that light is striking the eye along only one sight line. It is converging from numberless directions—after having diverged from one or more radiant sources and then reflecting and scattering off of objects in my environment. This apparently instantaneous build-up of converging, interlocking, and reverberating light rays, the whole of which envelops me, specifies my visual world. That is, it informs me of the world without my having to process flat, impoverished visual images.

Hence, for Gibson, ambient light implies information, but not isolated bits of information that, to become true or truly helpful information, must then be pieced together or coordinated by the brain. Rather, the information is ecological or relational by virtue of the ecological action of light. When, therefore, I look at a rock, say, I see it as it has been threaded together by multiple light rays. It is a particular object in my environment, but it is visually announced by ambient or surrounding light, not merely by light rays particular to my head-on alignment with it. And since ambient light is densely articulated by rays converging and intersecting from various directions, objects generally strike us as densely or deeply articulated. We may see only their frontal aspect, but we sense hidden depths or sides, because light rays reflecting from those hidden aspects play into the construction of ambient light and consequently into our perception of things.

One can see that Gibson's outlook departs radically from the traditional model of vision. That model cuts the visual field into pieces (isolated images) and then has the brain pick up the pieces so as to reconstitute the field as information. Gibson, however, realized that, owing to light's ecological action, the field is itself highly informative by virtue of its wide, deeply articulated nature. What is more, the field—that is, ambient light—leans into observers so as to undermine the Cartesian assumption of perfect mind-matter dissimilarity. To a significant degree, ambient light already includes—already informs—observation, and that precisely is Gibson's point and the basis for his contention that the standard theory populates the story of vision with fictitious elements. Because we are participatory with nature rather than aloof from it, the real story is more congruent with everyday experience and more affirming of its significance. Additionally, the real story is more likely to situate us in the world in ways that underscore our far-flung relationality with nature.

Invariance, Relationality, and the Speed of Light

Gibson's model situates us in the world and thereby nudges us toward a more relational view of things. A case in point is Gibson's claim that perceptual information is given by "a complex of relations" rather than by piecemeal inputs.[15] When considering the shifting of the optic array as one moves about, Gibson remarks that while different parts of the array move faster than other parts, no one should liken the parts to "separately moving" elements: "Optical motions . . . are a set of relations."[16] This much seems hardly worth stating. When landing an airplane, the pilot does not suppose that the approaching runway is moving faster than the lateral countryside, even though it appears that way. She perceives—she knows—that this appearance is a function of her motion relative to a particular strip of ground, and that apparent speed diminishes as lateral distance from the plane increases. All this is a set of relations, not a jostling of separately moving entities.

While this point may seem obvious, Gibson extracted from it two interrelated insights. First, sets of relations embody invariants from which we gain information about the world. "The active observer," said Gibson, "gets invariant perceptions despite varying sensations."[17] While sensations typically vary from moment to moment (particularly during movement), perception involves picking out stabilities that persist over time, whereby we lock on to lasting patterns in the environment. For example, while lifting off in a helicopter, we perceive—despite shrinking visual images—that trees maintain a constant size. This is because the apparent size of trees changes uniformly (invariantly) to let us know that their actual size is not changing (invariant). Something similar happens to colors of objects when those objects are exposed to varying lighting conditions. We do not perceive actual changes in color because we pick up on some invariant that runs through the gamut of changing color sensations. To follow Gibson: "The hypothesis is that constant perception depends on the ability of the individual to detect invariants, and that he ordinarily pays no attention whatever to the flux of changing sensations."[18] The implication here is that perception is an active, searching process. Already looped into nature, the physical senses actively "feel out" its invariants.

All this, of course, occurs involuntarily. It is a starting point built into our nature but one that prefigures the conscious quest for invariants, the scientific aim to discover the constants of nature. If Gibson is right, our perceptual systems attune us to invariants long before we take scientific cognizance of them. In science we hit upon them at a higher turn of the spiral, generally not realizing that earlier turns play into our own nature, so that pre-reflective perceptual experience is much more than passive, camera-like registration of the "outside" world. It is active (re)attunement to a world with which we have primordial complicity.

The second insight is a variation of the first, but while the first underscores invariance, the second highlights relationality. The two emphases, I believe, are different sides of the same coin. Gibson allows that in some explanatory contexts it may be appropriate to portray light as if it were composed of separately moving entities—photons, say. He points out, however, that this characterization does not carry through to perceptual experience. Recall the experience of landing an airplane. The pilot's optic array can be mathematically graphed with longer vectors representing faster-moving points in the array. But the vectors do not, of course, correspond to discrete elements because, as Gibson says, "Optical motions have no inertia and do not collide with one another."[19] Rightly understood, the vectors depict a set of relations. While this point was made above, it deserves reiteration for the simple reason that light has increasingly come to be seen as relational in modern physics. In one context, that relationality displays itself as photon entanglement: the way counter-propagating photons maintain instantaneous relation or contact despite arbitrarily large separation intervals. But if we were to match vectors to the photons, what would be the correct way to think about those vectors? As representing optical motions in an expanse of light, or as representing the motions of distinct and separately moving entities? The latter characterization plays to the sensibilities of classical physics while the former agrees with visual experience as described by Gibson.

My sense is that Gibson's view comes closer to the mark than classical physics, which sees the cosmos as a system of interacting parts. The classical model is not wholly incorrect, but it gives primacy to parts. They self-exist, or so it seems, and what they do next—in this case, interact—is secondary to the fact of their independent being. But Gibson shifts the emphasis toward interaction or relationality so that we get a sense of a system that is more than just an aggregate of parts. By having relations figure into the reality of the parts, Gibson distributes the possibility of particle interaction throughout the entire system and thereby prepares us for modern physics' revelation of entangled particles. It might be that particles and relations are mutually constitutive, and that is why particles remain entangled or related despite apparent separation. Like optical motions in the optic array, they cannot but hang together relationally.

Gibson, to my knowledge, never addressed the issue of particle entanglement. In unpublished musings, he did, however, take up the question of the speed of light. While he was generally content to speak of light rays and photons when describing light's behavior, he recognized these entities as theoretical abstractions remote from everyday experience. Moreover, he came to question the premise of their motion through space and time. The primitive experience of light, he noted, consists of its immediate availability. Although we may contrive situations where we wait for a light signal, the simple act of seeing affords us instantaneous access to the world. We open our eyes and the entire visual field is there. It does not enlarge over time. It does not take time for light rays or photons to move through space and inform us of the world.

This simple observation was problematic because physics conventionally portrayed light, the agency of vision, as moving through space and time. In Gibson's mind, however, this portrayal issued up from armchair reflection, not actual experience. Space, after all, has no visual properties. Accordingly, he insisted that "the concept of space has nothing to do with perception. Geometrical space is a pure abstraction. Outer space can be visualized but cannot be seen."[20] With no space to travel through, he reasoned, light requires no time to reach its destination, which is why Gibson further insisted that "information is not *transmitted* [and] the speed of light is irrelevant to vision."[21]

For Gibson, the *finite* speed of light was irrelevant. His research indicated that "[a]t the perceptual level illumination reaches a steady state *instantly* and [therefore] the speed of light considered as radiant energy is irrelevant."[22] But here he might have reflected that, according to special relativity, the speed of light is not unambiguously finite. Moreover, a parallelism exists between light in special relativity and Gibson's notion of ambient light: both are, in some routinely overlooked sense, unbounded by space and time. As suggested earlier, Gibson, in developing his concept of ambient light, gave up thinking of light-mediated images moving unidirectionally through space (along a single linear trajectory from stimulus to receptor). He proposed instead that specific visual information, though it may appear to reach the eye along a single sight line, arises from the entire optic array of one's experience. Just as bats do not hear isolated echoes and then weave various sensory strands together to develop a full picture of their environment (they respond to an ambient complex of interblending acoustical pressures—one that already incorporates their presence), we do not weave visual experience together from isolated images.[23] The experience, the full picture, is implicit in ambient light, and changes in ambient light (coinciding with changes in visual experience) occur instantaneously and holistically.

Objections

Light is so deeply ingrained in the seeing experience that its strangeness normally goes unremarked. The tendency, then, is to conclude that it, like the bodies it illuminates, moves through space and time. Against the contrary thesis (the thesis I am developing), the following objections may emerge.

(1) Some may object that all or much of the foregoing is just wordplay: to see an object, one perforce must not see everything that goes into the constitution or description of that object. If, for example, I could see the molecules that constitute my pencil, then I would not be able to see the yellow, cylindrical object lying on my desk. In brief, one level of vision—one visual texture—necessarily excludes another.

The argument is sound as stated, but it cannot be generalized to include light, for light per se has no visual texture whatsoever. By itself, it does not reside, like molecules, at a particular level of vision or means of detection; if it did, it would be as powerless as molecules to conduce vision at other levels. It would be another thing to be seen and not a principle of seeing.

(2) Another objection strikes an analogy between the delivery of light-mediated images and the delivery of other kinds of reports. I agree with Claire Michaels and Claudia Carello that when vision is modeled as the rectilinear, ray-like delivery of visual data, "a variety of intractable pseudoproblems emerge,"[24] and like them I am confident that this model self-destructs of its own inadequacy. In this case, the seeing experience is likened to the postman's delivery of a letter. The analogy might seem persuasive at first blush, but when scrutinized, it fails badly in three interrelated ways. First, the arriving postman is not invisible, and his delivery of the letter constitutes an event. By contrast, light's *delivery* of images (though not the images themselves) is clean and frictionless—eventless. Second, the postman may be hailed from a distance; he may be seen or heard as he approaches. Light, however, as a kind of *ur*-messenger, cannot be apprehended until it makes its delivery,[25] and then—to note the third failing—the so-called message is a *real* distant event, not a report of something else but simply something else. We do not (except

when speaking abstractly) talk about seeing images or visual reports of trees and buildings: we see trees and buildings. In one timeless stroke, unseen photonic immediacy opens out onto the seeing of distant bodies, and there is no experience of receiving messages or reading reports in order to imagine distant events. Distant events are there unbidden.

(3) Thirdly, one may protest that since blind people get by without light, and get by intelligently, it cannot be as important as I have made it out to be. This objection misses the larger point already noted: light is universal in the sense that its speed bounds the motion of material objects by playing into the way those objects negotiate space and time. Conceivably, our entire spacetime existence is bounded and structured by light in ways difficult to imagine. Gibson focused on the most obvious experience of light—seeing— and discovered what he felt was its atemporal, aspatial aspect. It does not follow, however, that those who do not see are outside light's universal embrace.

(4) Finally, there is a tendency among some to reject explanations of light such as this as "metaphorical" or "poetic." This stance, however alluring, erroneously assumes that metaphors can be avoided when speaking of such indeterminate entities as light, photons, electrons, and so on. The question to be asked is not whether a particular explanation is rooted in metaphor, but whether we realize that even our most hardheaded, seemingly non-poetic explanations are so rooted.[26] Thomas Carlyle's dictum that "[t]he coldest word was once a glowing new metaphor" applies to the conventional understanding of light, which has hardened around the now cooled metaphor of moving light.[27] This metaphor, Toulmin points out above, once glowed with the promise of talking about light in new ways; now, however, it is widely taken as literal fact, even though its limitations have been thrown into relief by recent advances in physics and by Gibson's ecological model of vision. Perhaps, as Niels Bohr once urged, it is time to reach for new metaphors when contemplating physical reality.[28]

According to modern cosmology, light is ontologically prior to space and time; according to special relativity, light is indifferent to space and time, at least within its own frame. But—and this is the wonder of it all—what is a spaceless, timeless frame? Unbounded by space and time, the photon's frame might well be indistinguishable from the circumambient light-sphere that seems to stretch on endlessly—or Gibson's ambient light that is immediately informative of the world. Suffice it to say, there are good reasons for supposing that the timeless, spaceless action of light not only conditions the motion of bodies in space and time: it also structures visual experience.

9

PRE-REFLECTIVE EXPERIENCE

The intellectual life of man consists almost wholly in
his substitution of a conceptual order for the perceptual
order in which his experience originally comes.

William James, *Some Problems in Philosophy:*
A Beginning of an Introduction to Philosophy

"The most incomprehensible fact about the universe is that it is comprehensible," said Einstein.[1] This statement captures his belief that the world is congenial to our curiosity about the world. Mind and cosmos are on the same page—otherwise each would be inaccessible to the other. Einstein never said this precisely and in fact resisted the implication that mind and cosmos ultimately dissolve into one another. All the same, he believed that humans have an innate capacity to understand the world and wrote that the prospect of theoretically grasping the world "beckoned as a liberation."[2] He wanted a perfect map or theory of reality.

But others have found the business of theory construction problematic. It seems to put a veil between nature and our firsthand experience thereof. Johann Wolfgang von Goethe, for example, felt this way. He argued that while the explanatory language of science accommodates human interests, it also distances us from nature: "We never sufficiently reflect that a language, strictly speaking, can only be symbolical and figurative, that it can never express things directly, but only, as it were, reflectedly."[3] Few would dispute this point, but some might ask if there is a way of practicing science without lapsing into objectifying language. Goethe would say there is a way of practicing science while learning from nature and not just about it. That is, nature is not mute; its manifest ability to affect us, to draw us into its orbit, is proof of its capacity to communicate truth in its own idiom.

Because Goethe made this point while taking aim at Newtonian optics, and because some of his ideas anticipate those to which we will soon turn, we pause briefly to consider his attitude toward light. "If the eye were not sunlike," he asked, "how could we perceive the light?"[4] His response: "The eye has light to thank for its existence. From undifferentiated tissue of the organism light calls forth an organ akin to itself; and so the eye is formed by light for light, that inner light might stream forth to meet outer light."[5] On the one hand, this sounds perfectly reasonable, at least when recast in scientific language: eyes have evolved by responding to environmental pressures and openings. But on the other hand, it implies an originary immersion in nature, and that militates against the belief that we can extract ourselves from nature to view it objectively.

Echoing Goethe's poetic prose, I am arguing that light speed constancy occurs as inner light (mind) streams forth to meet outer light (world). The indissoluble union of the two lights, already in place by the time science posits a difference between them, keeps us married to a single speed value no matter our state of motion. Thus we have a constant of nature that is inclusive of our own nature.

This implies that light speed constancy is hard to get at because it is pre-reflective or pre-theoretical. As such, it is the very kind of thing modern phenomenology has come to prize. According to Emmanuel Levinas, phenomenology seeks to substitute "a human world for the world as physicomathematical science represents it."[6] Not that scientific representations of reality are bad per se, but when accepted uncritically they can overpower and disenfranchise pre-reflective experience. Be that as it may, even the best representations or theories are never quite clean. "Reality divided by reason," wrote Harida Chaudhuri, "always leaves a remainder,"[7] a puzzle or paradox suggesting human involvement. Light speed constancy, I believe, is rooted in that remainder. It helps us see that ultimately we cannot decouple or divide ourselves from a reality that includes and involves us from the start.

Modern Phenomenology

While describing a truth seeker's experience in book 7 of his *Republic*, Plato wrote: "Clearly he would first see the sun and then reason about it."[8] But although Plato acknowledged that we routinely perceive things before we reason about them, he held that it is reason or analysis that puts us in touch with truth. So for Plato, as for many later thinkers, reason was primary, even though it *came after* perception and was therefore dependent on it. Owing to its emphasis on reason, this way of grasping the world, of trying to discern its *unseen* truths, is known as rationalism. Indeed, Plato tended to distrust sense experience and saw it as a kind of illusory veil to be penetrated by reason.

Opposite rationalism stands another broad epistemological tradition, empiricism. Believing that truth follows the flow of sense experience, empiricists incline toward observation and experiment. The results thereof give one something trustworthy to reason about; they guide the mind and keep it in check.

Modern science is generally deemed a healthy blend of both approaches. While sense experience may lead us astray, careful analysis and experiment keep the intellect honest, keep it from getting caught up in the delights of armchair reflection. But despite this productive blending of different traditions, some critics hold that both rationalism and empiricism embody a common set of self-defeating attitudes toward reality. Science therefore is similarly afflicted and thus unprepared to grasp elemental truths about the world.

According to these critics, those truths reside in phenomena, which science distances itself from in the interest of theory development. The word *phenomena* comes from a Greek cognate meaning "appearances" or perceptual manifestations involving such basic experiences as color, sound, touch, motion, and so on. Scientists like to step back from appearances in order to interpret them. This tendency to interpret and analyze phenomena, and to gather them into comprehensive theories, presupposes that they are apart from us and therefore fully amenable to outside scrutiny. It also assumes that they are mute or non-self-explaining—that is why human reason must bring forth their meaning and truth content. Even empiricism, which puts a high premium on sense experience, regards the raw experience—the phenomena—as uninformative until that experience can be intellectually processed. Phenomena are mere appearances until, through careful study, they can be turned into "facts," and facts are ideational truths.

The contrary view—phenomenology—recovers phenomena as the world's primary expression from which science and other second-order expressions have emerged. The phenomenal world was here long before people began to sift it into ideas, and the act of sifting it into ideas reaches back to the phenomena—the primary elements—themselves. Without them as primordial ground, philosophy and science would never emerge. Hence we cannot leverage phenomena onto a perfect map of reality; that is, a map that loops out of the world toward some non-worldly or non-phenomenal platform from which to see the world. What we can do, however, is try to wean ourselves from interpretative depictions in an effort to "get back to the things themselves." Students of phenomenology wish to recover or get close to an erstwhile pre-reflective or pre-analytic experience of the world.

Not that all phenomenologists think exactly alike. Edmund Husserl, widely regarded as the movement's modern founder, at times proposed that one could *fully* recover that pre-analytic experience. Others have disagreed, insisting nevertheless that much may be gained from partial recoveries. In this chapter I look at Husserl's doctrines of *Lebenswelt* and intentionality, each of which, he felt, recovers aspects of human experience glossed over and forgotten in the West. Getting back to those aspects—discovering them afresh after living with them since birth—reinforces points already made: (1) we are relationally tissued into the world; and (2) already so tissued, we cannot step back to get objective distance on nature's fundamental properties. They reach into our being as we reach out to grasp them. One result of this bi-directionality is constancy rather than variability borne of unilateral action on our part.

Husserl's *Lebenswelt*

Pioneered by Husserl in the early 1900s, phenomenology responds to, among other things, Descartes's famous dictum, "I think, therefore I am."[9] For Descartes this was a bedrock truth upon which the edifice of human understanding could be securely built. Or, to vary the metaphor, it was a self-luminous truth, a proposition whose blaze grows brighter when subjected to intellectual attack. To doubt it is to affirm it, for I cannot deny that I, in the very process of doubting it, am thinking. Therefore, indubitable proof of my existence lies in my thought, and, what is more, that is what I really am—a thinking being. I am not, at bottom, a physical being, because, unlike thinking or mind, body parts and sensations can be doubted. At this moment my hand looks and feels real, but I can find ways to doubt its reality without having that doubt flame up as self-evident truth. I may, for example, be dreaming that I am looking at my hand. Or, for all I know, I may have no hand at all but am merely hallucinating such. This seems very unlikely, but it is possible, according to Descartes, for, unlike the mind, nothing about the body indubitably secures its existence.

Descartes's views contributed to and fit into an emerging habit of thought that saw mind and body as disjunctive elements of human experience. Mind was active, knowing, and, according to Descartes, unextended in space or physically dimensionless. In a manner approaching God's existence, mind could not be readily mapped into the physical cosmos. The body, by contrast, was nothing but inert, unknowing physical substance ruled by mechanical necessity. While the mind acted, the body merely reacted. Thus the mind could be thought of as disembodied, at least in so far as it had little or nothing in common with the body. It surveyed and knew the world from within an unknowing body whose very existence could be questioned.

The mind, to speak more generally, was in the physical cosmos but not of it. Everything physical was deemed lifeless and consequently utterly different from, even alien to, the active, knowing, spatially unextended mind. Armed with this outlook, scientists could adopt the posture of the God-like spectator who magisterially surveys nature without participating in it. They could, moreover, come to a conclusive understanding of nature. As Descartes put it, the human edifice of knowledge could be put on an absolutely secure foundation and then erected without any faulty workmanship. This, of course, presupposed learning to reason correctly.

No doubt Descartes's argument in behalf of mind has been persuasive to millions, either as it was directly given or as it has sedimented into Western culture. Consider, however, that in proving his reality as a thinking being, Descartes limits himself to the only slice of human experience within which intellectual proof makes sense, and that is thought. A hunter-gatherer does not ask himself if he is real—that question is pre-answered by his everyday involvement in the world. But if he were to take up that question in a Cartesian manner—in a manner satisfying the demands of Western logic—he would have to think. And in doing that, he would acknowledge himself as a thinking being, which acknowledgement could be taken as proof that he exists, at least as a thinking being. So Descartes's proof seems to involve a certain circularity arising from the fact that proof as we define it is the special province of thought, and thought is transparent to itself.

What Husserl realized is that our everyday involvement in the world precedes and allows for philosophical analysis of the world. Eating, sleeping, breathing, getting a drink of water, throwing a rock—all these activities and many more constitute a pre-intellectual ground of being upon which thinkers such as Descartes build their intellectual edifices. So the supposedly bedrock truth of "I think, therefore I am" does not reach all the way down. It rests on a vaster, deeper bedrock, one so foundational and familiar as to be neglected by those who have learned the delights of abstract thought. Even when neglected, however, that deeper bedrock—what Husserl called the *Lebenswelt*, or life-world—never gets left behind. As Descartes developed his brilliant intellectual program for setting humanity on the path to truth, he did not transcend his need to eat, drink, visit the bathroom, and so on. His pre-intellectual life remained fully in force, all the while functioning as the forgotten or unnoticed ground of his philosophical ruminations.

Put another way, intellectual understanding is rooted in and propped up by intellectually unprocessed experience. Factual understanding reaches back to phenomena which "stand under" that understanding and which, in some ways, remain untouched by it. Living off the top of the *Lebenswelt* as they do, intellectual categories prevail only at a derivative level. They do not hold, they never arise, at the pre-intellectual level. For example, arriving at my office in the morning, I turn on the overhead light and adjust the blind. These actions are perfunctory, but if I think about them, I tend to regard them as separate events. The light switch, after all, is near the door, while the blind is on the opposite side of the room. But, as Husserl pointed out, actions always entail spheres of action with widely extended, indeterminate boundaries. Even if I close my eyes while flipping the light switch, I feel myself in a large setting. The floor pressing against my feet connects into and thereby "grounds" me to the earth, and the earth implies by contrast the overarching, outwardly stretching sky.

This large setting, moreover, imparts meaning to my actions. Turning on the light and adjusting the blind make no sense in and of themselves. They are performed with the immediate past and future in mind—I have had a good night's rest and want to get to work. Thus my sphere of action, aside from being spatially extended, also has a temporal aspect. To generalize, things are more widely and intimately interrelated than we normally suppose, at least as they show up in the *Lebenswelt*, the world of lived experience. That world hangs together because firsthand experience is hard to demarcate— to rule off in space and time—and therefore all but coincidental with the world itself.

The *Lebenswelt*, Husserl pointed out, "was always there for mankind . . . just as it continues its manner of being in the epoch of science."[10] Owing to its systematic unity—the way it interrelates globally despite undergoing local change here and there—and its indeterminate expansiveness, the *Lebenswelt* is the backdrop against which science plays itself out. "Backdrop" suggests stability or constancy, a kind of permanence that recedes from view so that impermanence or change may show up, and perhaps the obvious instance of such is a movie screen. The white homogeneity of the screen—its relative blankness—affords the play of shapes and colors that attract human interest, and those ever-shifting images may seem the very picture of visual autonomy: the presentation apparently begins and ends with their showing. But, of course, this attitude stares past the screen, whose featureless constancy functions as a kind of visual and mental absence—a kind of forgetfulness—against which images show up and appear to assume a reality all their own.

We can take the analogy a little further. During the movie, the play of variegated images may be so compelling as to cover over any hint of unifying background constancy. Nevertheless, someone like Husserl, upon trying to experience those images as simple, unprocessed phenomena, might begin to pick up on far flung articulations and constancies normally overlooked. Aside from remarking on the seamless unity of the *Lebenswelt*, Husserl observed invariances or constancies that persist throughout perceptual experience. The most fundamental of these is that the *Lebenswelt* is "not itself relative,"[11] meaning that it always bulks omnipresently large even as we try to step away from it. Analysis does not put it in its place, so to speak, relative to some new, outside perspective, for any perspective on it is always from within. This, said Husserl, is because the *Lebenswelt* is pre-given: by the time we get around to giving it philosophical consideration, it has long since arrived. More precisely, it has long since shaped us ontologically, and so every instance of trying to step away from it becomes another of its many expressions.

In a word, the *Lebenswelt* is horizonally all-inclusive. Trying to step outside it is like trying to step beyond the horizon. This simile, now familiar, subverts the well-worn Cartesian assumption that we, owing to our radical dissimilarity with nature, may view it from an outside perspective. But phenomenologically speaking, we never get beyond the center of the world: we—you and I—are the pivot about which the world swings, even as we adopt intellectual postures pretending otherwise. Attending to firsthand experience, Husserl discovered an ineluctable centrality implying inescapable involvement in the world. Each end of the stick—observer and observed—constitutes the other, and this bilateral interpenetration or mutually constituted affinity spontaneously opens out as everyday experience.

This opening-out, however, is not without limits or horizons. Husserl's *Lebenswelt* may be regarded as an openness with intrinsic, mutually implicative possibilities and impossibilities. We cannot step beyond it because it is constitutive of our being, but this fact ensures that it remains open and living. Unlike a prison cell whose fixed walls resist our movement, the circumambient *Lebenswelt* does not confine us: it shifts with us because, in some elemental sense, it is us. Its horizons are our own, experiential arcs of meaning where our perception interblends with things physically other—trees, buildings, mountain ranges, stars, and so forth.

Intentionality

Grasping Husserl's doctrine of *Lebenswelt* prompts the following realization: taken merely as physical things, we are small parts of the cosmic whole, but taken as perceptual beings, we are "unbounded portions of the whole."[12] We have wide identity with other things, at least in so far as they disclose themselves to us. As Robert Sokolowski puts it, we are "datives of manifestation."[13] By nature we are open to the manifestation of all sorts of things, and this openness coincides with the openness of the world.

Implicit in all this is Husserl's concept of intentionality: consciousness invariably entails consciousness of something—a rock, a tree, even an imaginary, physically impossible entity like a round square. Put another way, mind alone is inoperative or unmindful; it breaks into being by intending or targeting things other than itself. Contra Descartes, then, it is not as if there is a sharp break between world and mind. The two interpenetrate, and so the Cartesian image of a knowing mind peering out onto an inert, uncomprehending world gives way to a sense of mind-world symbiosis. With nothing "out there" to intend, consciousness with its various feelings, thoughts, and moods goes unaccomplished.

Intentionality implies relationality, and, in this case, a relationality originative of being rather than parasitic on it. Mind and world interrelate or interlock with a click, the click being existence as we know it. We come back, then, to a point made earlier while discussing special relativity. We noted that physical objects do not possess intrinsic length and time values; that is, values that prevail in the absence of observers or in the presence of some privileged observer. There are no privileged observers, and there are no "true" or universal default values that objects assume when unobserved. The precise length of an umbrella, say, is given as the umbrella is interactively coupled with—measured by—an observer.

Special relativity, in other words, does not sanction the tendency to regard the umbrella as having a standard, observer-independent length value that then shows up or appears differently in varying situations. Instead, the appearances—the phenomena—go all the way down; they are not "mere" appearances but rather experiential truths borne of one's relationship with objects of the world. Grasping this point is tantamount to grasping the similar phenomenological point that intentionality is not something added to consciousness: it is consciousness. Relationality or intentionality is there from the first and is the means by which the contents of consciousness (intended objects) effect consciousness.

To be sure, this may all seem highly counterintuitive. Most of us are comfortable with the mind-world dichotomy, even if we cannot justify it philosophically. But, as Husserl pointed out, the dichotomy is problematic. Looking about, I see a tree. In which Cartesian domain—mind or world—does the tree exist? My normal inclination is to say that it exists primally or objectively in the world and then shows up as an idea in my mind. But there are difficulties with this account. For one thing, it seems uneconomical to count the tree twice, first as an object and then as an idea. A related difficulty turns on the question of how to compare idea with object when neither is distinct from the other. I cannot lay the two side by side to see if my idea of a tree accurately resembles the tree itself, now apprehended as an object. This, of course, is because I cannot step outside my consciousness of a tree—what Descartes would call my idea of a tree—to encounter the tree again as an idea-free object.

Husserl got out from under the mind-world or idea-object dichotomy by insisting that intentionality is primary, not something that occurs in the aftermath of mind and world. The dichotomy, and the question it prompts regarding the whereabouts of objects, arises only if we assume that the mind is a kind of self-bounded oasis of thought in an alien, unthinking desert. But intentionality un-bounds the mind, or helps us to see that it was never bounded to begin with. To assume, then, that something is "in here" as an independent idea or "out there" as an independent object is to *add to* what is directly given, and it was precisely these kinds of additions that Husserl was trying to avoid. Further, from the point of view of pre-reflective experience, the tree is given only once and given immediately. Why then develop a story about the tree producing an optical effect that, when taken in by the brain, gets translated into an idea or image of the tree? This story is another addition to experience, for we never perceive the production of an optical effect and its subsequent translation into an idea. Things never build up in this manner; perceptual experience is clean of these supposedly ancillary processes.

This cleanness, this economy of action in the face of theoretical addition and complication, deeply impressed Husserl. It was as if nature is more parsimonious than science and philosophy had bargained for. Theorists, while subscribing to the ideal that nature behaves economically, seemingly could not fully tolerate its vast elegance—hence their theoretical interpolations, their complications in the face of parsimony. Of course these insertions were made to explain the world, but they embodied assumptions that, again, are not directly given to consciousness. As noted, before the *Lebenswelt* gets characterized as a pluralistic aggregation of separable objects, it is experienced as a seamless, all-inclusive unity. Not only that, but as Husserl further pointed out, iterations of unity—of economy— inform our experience of putatively separable objects. For Husserl this was another aspect of intentionality.

We catch snatches of economy with every perceptual act. Glancing at a book, I apprehend or intuit it as seen from other angles of vision—that is why it shapes up as something substantial, something more than a façade or illusion. It stands forth as a solid, unitary object with hidden (yet to be seen) depths or aspects, an object that will "survive" any number of visual transformations. These transformations are enacted as I rotate the book and open it up, but if they were not implicit in my initial glance, each might well come off as a disjointed experience. Many different books would then displace the optimal economy of a single book. This, however, does not happen. The book is invariant or constant across any imaginable visual transformation.

What is being underscored is the unitary nature of things, both as things expansively cohere as a system and as single, seemingly self-standing things quietly articulate into the system. And not to be forgotten is the perceiver's widely ramified and already implicated presence. As we will see in the next chapter, Maurice Merleau-Ponty extends this analysis and relates it back to the physical body. Far from being an inert appendage to the knowing mind, the body for Merleau-Ponty is a sensing, experiencing membrane already tissued into the world. To be sure, its scientific properties resemble those of other physical objects in the world. In terms of experience, however, it is profoundly different. I can step away from the piano in my living room but I cannot step away from my body. I cannot in any firsthand way gain objective distance on it. It centers me in the world, gives me a first-person vantage point from which to affect second-person and third-person attitudes toward the world. It expresses my elemental inclusion in the *Lebenswelt*, the immediacy of life experience that cannot be reduced to or overtaken by theoretical propositions. In these respects, my body is unique among objects in the world. For me it lives; it stays in play as experience. This is a claim we have already made about light.

10

BODY, WORLD, AND LIGHT

*If we look at distant stars, then our minds stretch out
over astronomical distances to touch these heavenly
bodies. Subject and object are indeed confused.*

Rupert Sheldrake, *Seven Experiments That Could Change
the World: A Do-It-Yourself Guide to Revolutionary Science*

Often linked with James Gibson, whom he may have influenced,
Maurice Merleau-Ponty sought to describe perceptual experience
while eschewing talk of sense data, photons, retinas, and rods and
cones ("with only myself to consult, I can know nothing about this").[1]
His insights underscore subject-object and object-object interrelations, the various ways the world hangs together as a unity despite
the analytic tendency to split it into parts. His slogan might well
be, "What perception brings together, let no mind part asunder,
for mind or analysis lives off the top of perceptual experience." Going further, our complicity in things keeps us from rising above
the flux to take theoretical snapshots of the world. Thus, built into
Merleau-Ponty's outlook is an admission of incompleteness or indeterminancy, a recognition that the game of reason is a game of
catch-up.

In science there is no better illustration of the catch-up principle—nature's capacity to stay ahead of inquiring minds by already including them—than Einstein's second postulate. No motion or mental maneuver on our part alters the speed of light, evidently because light speed constancy marks our elemental inclusion in the cosmos. Constitutionally knit into the system—Merleau-Ponty would say incarnated in the world—we cannot independently overtake it. Thus while light brings visual and intellectual clarity and thereby abets the urge to map and analyze, it ultimately stays in play only as experience. To reiterate a central argument, light is not distinct from our experience thereof, and attempts to abstract it from experience by mapping it into separable, self-contained parts have proven unsatisfactory. When tested against reality, such mappings implode upon the novel possibility of particle nonseparability or quantum entanglement. Had Merleau-Ponty lived long enough to witness this development and had he taken an interest in it, he would not have been surprised. Years earlier he was thinking in the key of nonseparability upon contemplating phenomena sans theoretical overlays. This is particularly true of visual phenomena.

What Perception Gathers Together

In his *Phenomenology of Perception*, Merleau-Ponty asks why objects in space immediately show up as having depth or thickness, as being three-dimensional, when, according to the standard scientific model of vision, their visual images are two-dimensional. Trees, for example, are not flat; even when viewed from a single perspective, they have a kind of perceptual heft that goes beyond the claim of two-dimensionality. Or, looking at a house, I know that it is more than a façade—indeed the question of whether it is a façade normally never even registers. The unseen but still experienced thickness of objects, in short, is a full-blown aspect of visual perception, immediate and unsolicited.

Vision science is helpful here, but only partly so. To be sure, convergence of sight lines from both eyes affords depth cues, as do other mechanisms such as accommodation (the differential shaping of the focal lens for objects at varying distances). But, as just noted, Merleau-Ponty wishes to address the visual experience *before* theoretical elements—elements not showing up in the experience— begin to settle in. What is more, he is talking not about depth perception per se but about the sense that objects possess hidden depth or substance that nonetheless gets partly unhidden in the moment of seeing—hence the *sense* of hiddenness. An object's depth, in other words, while not apparent or visibly given in the way visual data are given to a camera, is yet implicit in the visual experience.

Like Husserl, Merleau-Ponty insists that depth arises from the interplay of focal figure (the object of interest) and visual background or field. Although centered on the focal figure, that background is populated by other objects, each of which has a different angle of vision, so to speak, on the figure. These other angles of vision get folded into the viewer's angle so that she implicitly sees the object from multiple perspectives. The example Merleau-Ponty offers is that of looking at a lamp on a table. I see the lamp from my angle of vision, but I see it as it is "shot through from all sides" by many other perspectives originating with objects in my visual field[2]—the fireplace, the walls, the bedpost, and so on. That is, the visual field is an indivisible whole: visually speaking, all parts of it reside potentially or actually in any single part. When I focus on an object, my residence therein, so to speak, opens hither regions of sight which, in turn, direct my gaze back to the focal object, so that I "see" it from many angles. Thus, as Merleau-Ponty puts it, "every object is the mirror of all others."[3]

Put another way, the focal figure gets rounded out by surrounding objects in the visual field, and this occurs because those objects mark various points in space from which the focal figure may be seen. To follow Merleau-Ponty: "Any seeing of an object by me is instantaneously reiterated among all those objects in the world which are apprehended as co-existent, because each of them is all that the others 'see' of it. . . . The completed object is translucent, being shot through from all sides by an infinite number of present scrutinies which intersect in its depths leaving nothing hidden."[4]

Of course the house is not translucent in the sense that one can see through it. The viewer, however, feels that it is more than a façade owing to its "being seen" by surrounding objects, each of which implies an alternative angle of vision.

This analysis, of course, departs from the traditional scientific model of vision that posits the rectilinear transfer of sense data from object to subject. On that account, sense data from simultaneously seen objects converge independently on the eye along paths defined by the different angular positions of the objects relative to the viewer. Since each path of travel is unique to a particular object (or part of an object), there is no reason, or even allowance, to assume, à la Merleau-Ponty, that the various objects interact visually among themselves while interacting with their viewer. While the convergence of incoming sense data may be tightly packed, tiny angular displacements ensure that it is not interactive.

But for Merleau-Ponty all this is theoretical. Sense data moving through space along geometric paths: no one sees either the data or the paths. They are, as Rom Harré puts it, "not seen in nature . . . [but] drawn on paper."[5] What one sees are three-dimensional objects in space. Or, to give a more elemental description, one sees those objects simultaneously. They are together—caught within a single perceptual vista—and that togetherness implied for Merleau-Ponty a primordial unity that he equated with space. He consequently rejected the standard model of space as a neutral setting for the location of objects or events. Within this model, space has no cohesive power, and so when several things are seen simultaneously, that simultaneity is generally deemed a mere consequence of our wide-angle visual field. Thus our ideas about space diverge from our visual experience thereof. Visually speaking, space is an expanse simultaneously populated by myriad things. Conceptually, however, that unitary expanse is iterated out of sight by those things, each of which may show up as a distinct entity trembling on the possibility of a spatially separated, self-contained object. Space consequently comes to seem a matter more of separation and interval than of simultaneity and togetherness—more a disjunctive relation than Merleau-Ponty's primordial or pre-reflective conjunctive relation.

Departing from the standard outlook, Merleau-Ponty insisted that "instead of imagining [space] as a sort of ether in which all things float . . . we must think of it as the universal power enabling them to be connected."[6] Our most expansive sense—and hence the one most intimately associated with space—is sight, and sight is an openness wherein things show up together: "Vision alone makes us

learn that beings that are different, 'exterior,' foreign to one another
are absolutely *together*, are 'simultaneity.'"[7] And elsewhere: "vision is
tele-vision, transcendence, crystallization of the impossible."[8] What
is more, sight is a *self-inclusionary* openness—we are already gathered
into this simultaneity: "We must take literally what vision teaches:
namely, that through it we come in contact with the sun and the
stars, that we are everywhere at once."[9] And to go one step further,
this "contact" is a "gift" or given prior to any analysis of vision:

> To say that I have a visual field is to say that by reason of my
> position I have access to and an opening upon a system of
> beings, visible beings, that these are at the disposal of my gaze
> in virtue of a kind of primordial contract and through a gift of
> nature, with no effort made on my part; from which it follows
> that vision is prepersonal.[10]

What Merleau-Ponty means is that although vision may be
personalized or particularized to single lines of sight, its expansive,
unitary character precedes such particularization and remains in
play despite it. At the far end of the line of sight, so to speak, is
the wide-open visual experience. And the widely inclusionary nature of that experience implied for Merleau-Ponty a very different
kind of space than that idealized by science, wherein things may be
viewed dispassionately. He thus stated that "our body is primarily
not in space; it is of it."[11] Space is indistinguishable from the body's
materiality.

As noted in chapter 1, relativity theory has prompted similar
thinking. Milič Čapek wrote that, thanks to Einstein, it is now correct "to speak of space being fused with its changing and dynamic
physical content."[12] As material realizations of spacetime, physical
bodies are imprecisely bounded portions of the spacetime expanse;
spacetime does not begin where they end. Rather spacetime and
physical objects mutually condition each other so that no clear line
can be drawn between the two. This mutual impingement, I suggested earlier, makes plausible Einstein's proposition of an absolute
speed limit. Whereas Newton imagined space as a physically featureless expanse offering no resistance to the motion of physical
bodies, Einstein saw it as deeply informative of those bodies and
therefore responsive to their motion. It would react back on them to

conserve the elemental or global properties of the system, and light speed constancy issues up from that conservation. This constancy, then, has nothing to do with any particular thing *in* the cosmos; it is *of* the cosmos and therefore expressive of its fundamental nature. To offer a simple analogy, it is like the constancy of Boyle's law, which marks a global stability despite local fluctuations in volume, molecular kinetic energy, and so on. That is, the constancy arises from the systemic or universal nature of the system, and cannot be grasped by focusing exclusively on local entities such as single molecules or even populations of molecules.

Thinking in a similar register, Merleau-Ponty proposed that we are visually fused or "tangled up" with the world.[13] Change or difference is eye-catching, but it shows up against a background of sameness—a constancy—that is both easy to overlook and indicative of some elemental commonality whereby things get primally inter-situated. Our relation to the world, he stated, is a "thick identity . . . which truly contains difference."[14]

If things were perfectly homogenous, the thought of mind-world duality or difference would never occur, of course, and so we cannot say that the thick identity is exhaustive. All the same, it is more extensive than most people assume. Merleau-Ponty faults science for adopting the posture of the remote, disinterested spectator, which posture flows from the naïve belief that the mind is a closed system in full possession of its contents. On the contrary and as Husserl proposed, mind and world are compounded together. This compounding denotes difference—because there are two things compounded—while also signifying an originary, unifying congeniality that permits the compounding in the first place. Not only that, but the compounding or interpenetration turns the mind into an open system. Just as relativity theory blurs the line between spacetime and physical bodies, Merleau-Ponty blurs the line between the mind and perceived objects. In neither case can one strictly demarcate material bodies from their observers. Or perhaps better, in neither case can one think of either category—objects or observers—as self-confined. Each merges into the other to accomplish its own reality.

In relativity theory, the merging is enabled by shared spacetime. For Merleau-Ponty, it is perceptually witnessed in the way space expansively gathers things together, although, as in relativity theory, space is not separate from the objects populating it. Instead it is a conjunctive relation denoting object-object and object-subject togetherness, and this relationality arises from the fact that "my body is made of the same flesh as the world . . . the world [therefore] *reflects* it, encroaches upon it and it encroaches upon the world . . . they are in a relation of transgression or of overlapping."[15]

Four hundred years ago, Galileo, Newton, and others broke down the Aristotelian dichotomy between the corruptible earth and the incorruptible heavens. The two spheres were previously thought to be composed of different substances and governed by different laws. The collapse of this dichotomy, however, stopped short of human mind, which Descartes protected behind a veil of immateriality. Suspicious of Descartes's move, Merleau-Ponty let the barrier fall between mind and body and thereby between mind and world. What resulted from this leveling operation is the realization that intelligence, the ability to make sense or meaning of things, is not the sole preserve of a dimensionless point called mind somewhere in the brain. Rather, intelligence is already prefigured into and distributed throughout the whole, and by this means we have cognizance of and commerce with the world.

Touch-like Seeing

Although Merleau-Ponty's outlook runs against the grain of science, particularly as science has parsed the world into things that now strike us as different, self-standing objects, it is yet sanctioned by deeply familiar ways of talking, which acknowledge that one's being can vastly exceed the physical limits of one's body. When Abraham Lincoln spoke at Gettysburg, his presence was felt and witnessed by those in attendance. We normally explain this by saying that while his body remained a constant size, his persuasive influence radiated outward to inspire sympathy and courage. But this does not resolve the issue, for having said this, most would still want to confine Lincoln's *actual* presence or being to his body or, more precisely, to his body-enclosed mind. Such a stance, however, would raise the following question for Merleau-Ponty: by what means was Lincoln, confined to his body, able to profoundly alter the experience of other bodies or beings within his perceptual reach? This question takes on increasing urgency when we consider that space, as routinely conceptualized, has little or nothing in common with physical beings and thereby separates them from each other. That is, since space was already there before beings came along, it exists independently of them and for that reason allows little if any commerce among them. How then did Lincoln overleap this alien and alienating element to charge the atmosphere with his presence?

Merleau-Ponty's response is illuminating. Space is part of the physiology of the world, not a pre-existing alien element or a neutral setting into which the world is dropped. Also sharing in this physiology, we are widely tissued into the world. The body is a kind of knot at which various strands converge, but the incoming strands also reach outward to give us perceptual awareness of things beyond the body. That is why someone like Lincoln can fill a large spatial volume with his presence. And that is why our perceptual or experiential presence—our being—is so much larger than our physical bodies. In the following verse Wallace Stevens adopts this point of view:

> I measure myself
> Against a tall tree.
> I find that I am much taller,
> For I reach right up to the sun,
> With my eye.[16]

Thinking in the key of a shared physiology, Merleau-Ponty would say that the eye touches the sun. The spatial interval drops out of sight, as it were, to bring subject and object into what he called the "strange adhesion of seer and the visible."[17] Here again there is a bivalence or ambiguity to light that both nullifies and sanctions the proposition under consideration—in this case, that visible objects are in direct contact with observers.[18] On the one hand, objects recede in size according to geometrical laws as they get farther away, and this would seem to clinch the matter: distance or interval keeps eye and object apart. But on the other hand, distance or interval per se has no visual texture, which is why it must be inferred from the overall context. Right now my rocking chair is about three yards away, but I "see" those three yards only in virtue of seeing other things with definite visual properties. Still more striking, those three yards do not delay or temporally stretch out my visual experience of the rocking chair. Even if the chair were three million yards away (and my eyesight were very keen), I would see it immediately upon turning toward it. After all, I see much more distant stars the very instant I look skyward. Of course I may explain this by saying that I don't really see stars or chairs or any distant objects at all; I see eye-impinging photons that convey images of these things. But, here again, a theoretical addition has been smuggled in, for there is no way to see photons apart from the images they supposedly convey, and, what is more, I never feel like I am seeing mere images of things. I feel that I am seeing things themselves, directly. And so for Merleau-Ponty the *pre-theoretical* visual experience consisted of undelayed, unobstructed contact with visible things. That is to say, interval-less adhesion with those things.

We have already noted that, according to special relativity, two answers are given to the question, "How long does light from a distant event take to reach my eyes?" In a straightforward sense light, having a finite velocity, covers the distance in a finite period of time. But in a more esoteric sense, light, in Einstein's phrase, plays the role of "an infinitely great velocity" and therefore takes no time at all. Here the key word is "role." It is not a scientific fact that light moves infinitely fast, and light has never been measured as moving that fast. But when Merleau-Ponty insists that "we are everywhere at once" in virtue of the expansive visual experience,[19] he is thinking in a pre-reflective, pre-measurement mode. He is recovering pre-scientific visual experience. "The first time we see *light*," wrote William James, "we *are* it rather than see it."[20] This is the Zen of seeing, what Merleau-Ponty called the "delirium" of seeing[21] and light's "*imaginaire*,"[22] and it is where the role of light as an "infinitely great velocity" or instantaneous velocity figures in.

The Role of Light; the Interval and Its Revocation

Light moves at constant finite velocity, according to science. In a pre-reflective sense, however, it is immediate, even instantaneous—that is, it effects immediate visual experience. The linchpin that holds these apparently contradictory claims together is light speed constancy, which, I have argued, denotes our inescapable involvement or pre-involvement in the world. To measure the speed of light is to measure something about the way we ourselves are measured or blended into the cosmos, and that universal blending pre-decides our measurement of light speed in favor of a universal value.

Or to give this thought an optical emphasis consistent with Merleau-Ponty's outlook, the theoretical revelation of light speed constancy occurs within the larger, more immediate, and fully unsolicited revelation of everyday visual experience. In this latter revelation, light's constancy subsists in its unfailing hereness or immediacy, its inability to be seen except while striking the retinas. At a higher turn of the spiral, this constancy becomes light speed constancy—the realization that light is its own yardstick. It makes sense that at some point in our quest to measure reality we should encounter something that cannot be conventionally measured, if only because at that point measurement becomes tautological. The elemental yardstick gets laid against itself, always returning a self-consistent, self-constant value.

Although Merleau-Ponty did not address light speed constancy, he did note that "the scientific critique of the forms of space and time in non-Euclidean geometry and relativity physics have taught us to break with the common notion of a space and time without reference to the observer's situation."[23] He further insisted that perceptual constancies arise from our pre-involvement with things perceived, and faithful to this notion, he tended to speak of lighting rather than light, realizing that lighting connotes the interplay of light, object, and observer. Speaking of color constancy, he wrote: "The lighting is not on the side of the object, it is what we assume, what we take as the norm. . . . it is anterior to the distinction between colours and luminosities."[24] Put differently, invariance across perceptual transformation signifies a prior complicity with the phenomena, one that subsequently cannot be unraveled or atomized into distinct color experiences.

The point he wishes to make is that such constancy implies a world with which we are pre-conceptually cognate; there is, therefore, an intelligence or pre-intelligence to our seeing that answers to the way we negotiate and experience an illuminated environment: "The lighting directs my gaze and causes me to see the object, so that in a sense it *knows* and *sees* the object."[25] Elsewhere he describes light or lighting as something that "remain[s] in the background . . . and *lead*[s] our gaze instead of arresting it."[26]

Light, in other words, facilitates seeing in unseen ways, having settled into our nature from the very start. That is why, for Merleau-Ponty, it is not "an absolute invisible, which would have nothing to do with the visible. Rather it is the invisible *of* this world, that which inhabits this world, sustains it, and renders its visible."[27] Already figured into the world, we are "naturally destined" to experience its overall meaning and figuration, and this, says Merleau Ponty, is the primary visual experience light affords us: the pre-existing harmonies and lines of force that, without any thought or effort on our part, visually blend us into the world, whereupon we may afterward take note of (apparently) solitary objects and processes.[28]

This conviction—that we are cognate with the world—motivated Merleau-Ponty's claim of direct-contact or touch-like vision. Having subtracted out unexperienced theoretical interventions such as photons, sound waves, and light waves, he found that perceived objects do not exist apart from a "reciprocal insertion and intertwining"[29] of object and observer. And so, perceptually coincidental with myriad perceived objects, we are expansively distributed throughout the world: "We must take literally what vision teaches: namely, that through it we come in contact with the sun and the stars, that we are everywhere at once."[30]

This claim, as absurd as it might sound, is not wholly beyond the pale of science. With the advent of relativity theory and the accompanying realization that no information or causal influence can outrace light, physicists were able to characterize reality as a light cone. Such is depicted below. The apex of the cone is a particular spacetime point from which a spark expands to constitute the cone. As the illustration indicates, the cone's horizontal aspect represents its expansion in space, while its vertical aspect marks its movement through time. For an observer at the origin, the point at which the spark originates, the cone depicts knowable reality. Its edges coincide with the speed of light; observers cannot see or know of events beyond that boundary. Routinely called a horizon, this boundary

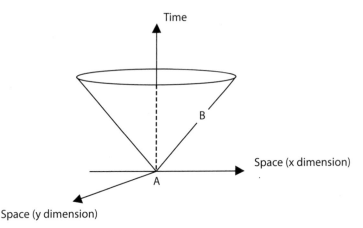

Figure 10.1. A spark marked event A expands to form a light cone. From a spacetime perspective, event B, lying at the edge of the cone and therefore linked by a ray of light to A, enjoys light-like or zero-interval separation from A.

is an epistemological limit arising from the ontological fact that no signal can outrace light.

Given Einstein's postulate that light is the fastest possible velocity, this is all quite straightforward. With light as the cosmic speed limit, the cosmos is visually mediated at light speed or less; we must wait upon light, as it were, for a picture of reality. But, as already suggested, this is not the complete story. In some sense we do not wait upon light because we are already light-infused, and so there is an aspect of simultaneity with regard to our experience of light that Einstein and Merleau-Ponty bring forward, albeit in different ways. Following Einstein (more on Merleau-Ponty shortly), the geometry of light is such that two events connected by a light signal are co-incidental in spacetime. That is, no spacetime interval separates them, even though an observer may deem them to be at different spatial locations. Two such events lie on the cone's edge and are said to enjoy light-like separation, which from a spacetime perspective means zero-interval separation.

Following Edwin Taylor and John Wheeler, we mark these events A and B in the illustration. After noting that in spacetime geometry the time factor is subtracted from the three space factors (each representing a different spatial dimension), Taylor and

Wheeler go on to state that *"The interval vanishes when the time part of the separation between A and B is identical in magnitude to the space part of the separation"* (original emphasis).[31] That is, when the two events lie on a line that bisects the space and time axes of the light cone, which line marks the cone's edge and, of course, light moving at its characteristic speed.

Taylor and Wheeler ask: "What is the physical interpretation of this condition?"[32] They respond by acknowledging that while the two events are not spatially coincidental, light nevertheless "travels one meter of distance in one meter of light-travel time."[33] Light, in other words, intrinsically covers one meter of distance in one meter of time; it is the trajectory at which space and time perfectly offset each other and thereby cancel out. Accordingly, *"The interval between two events is zero when they can be connected by one light ray"* (original emphasis).[34]

It would seem that, at some level, light-mediated vision might be party to this zero-interval immediacy. John Schumacher alerts us to one way this could happen by noting that "the light constituting the boundary of the light cone is *of* the cone, but not *in* it."[35] Of course light illuminates events within the cone, objects or processes moving less swiftly than light, but upon giving us visual witness of slower-than-light reality, light also brings us into its own economy of motion, or what Schumacher calls light's "order of movement."[36]

This order of movement is utterly different from that of sound, say, where acoustical events may be plotted against a backdrop of visual experience and thereby witnessed or at least anticipated before they arrive. Upon seeing a flash of lightning, we know thunder is on its way. But, says Schumacher, there is no comparable backdrop for visual events. We cannot see them or anticipate their coming; we just see them upon their arrival. In Schumacher's words: "Any truly limit movement must occupy a unique place in our experience: we can have no news of its upcoming arrival until it arrives itself, but then it has already arrived."[37]

What is the import of this fact? For one thing, it affords insight into the puzzling postulate that the speed of light is constant for all observers. "With no warning of the light that arrives at our place, we cannot resolve its movement in experience," says Schumacher.[38] Unable to see light from afar or to step back from it to view it objectively, we are locked into its unfailing presentness, and there is no spacetime backdrop against which its speed, always a matter of arrival for any observer, can be differently parameterized for differently moving observers.

If anything, light is its own backdrop, and so to "see" light is to see something already iterated out of sight by its perpetually unbanked reality. Light is of the light cone but not in it, says Schumacher; that is, not subordinate to the spacetime conditions it imposes on things within the light cone. And upon its arrival, light is here touching our eyes, stretching them out along the edge of the light cone while visibly announcing slower-than-light events in the cone. Inasmuch as "zero-interval linkages between events near and far" occur along that edge,[39] we might regard light's inevitable contact with or touching of the eye as one such linkage, and one whereby our seeing is assimilated into light's zero-interval economy. Of course this assimilation is not something we see, but it does enable the expansive visual experience.

Parting Thoughts

The foregoing is meant to affirm that light is shot through with ambiguity, and the same is true of light-mediated visual experience. In the case of relativistic or Einsteinian light, says Schumacher, it "alone can resolve numerous ambiguities in voice and sound, but cannot resolve its own ambiguities."[40] This is because as a limit movement light cannot be held up against any backdrop but itself, and so to "see" light is to see something already iterated out of sight by its own open-ended, ongoing reality. Thus light is a formless clarity or openness that permits seeing but is not overtaken by it.

Of course we cannot overtake light as we conventionally overtake fellow pedestrians or moving vehicles—that point has already been made many times. Neither, however, can we catch up with it perceptually, and that, I am proposing, is because it is profoundly informative of our being. Consider that water and air, two substances necessary for life, are colorless or transparent. Science makes little of this fact, but Merleau-Ponty might attribute it to the body's inability to fully descry elements deeply constitutive of its own fleshy being.[41] We are, after all, 70 percent water, and air, or its components, is distributed in one way or another to virtually every part of the body. Consisting largely of air and water, we tend to see through them upon encountering them in realms beyond the body. This is to suggest that water and air lack color or visual distinctiveness because they are deeply informative of our physiology, because they are inscribed into our flesh.[42] To some imperfect degree they are like T. S. Eliot's music that is "heard so deeply that it is not heard at all."[43]

The same and more might be said of light, whose transparency amounts to a perfect visual clarity or invisibility. We can at least find water and air outside us—at a distance—but not so light. To find light is to find something that tilts our vision toward things (material bodies) other than light, and only as we try to see light by its own dynamic do we begin to realize that it is not fully scaled into the metric of space and time. A little reflection, however, suggests that the universe, if it is to register as a single, unitary system, must be integrated by a different metric than that of pluralistic difference, which, after all, would add up to mere aggregation of parts rather than cosmic unity. What makes this view difficult to accept is our inability to see light's integrative embrace, but such is consistent with the whole meaning of light. By not announcing itself, the circumambient light-sphere goes on endlessly, never to be overtaken by sight and thereby caught within the kind of visual limit that marks the seeing of finite material bodies—that is, mere parts of the cosmic system. Light integrates reality by not showing up as another thing, another part, to be integrated.

This implies, among other things, that scientific light—what we might call theoretical light—cannot be uncoupled from our everyday experience of light: we are already integrated into a light-articulated cosmos by the time we decide to isolate and analyze light. Some, of course, will insist otherwise, but the pre-conceptual mystery of togetherness, simultaneity, or ubiquity ("everywhere at once," as Merleau-Ponty puts it) would yet persist in the wake of this first fundamental error—that of detaching light from our experience thereof. Listen again to Arthur Zajonc, a physicist with phenomenological leanings. By talking only of light, he treats it as a separable entity, but then concludes that it is profoundly non-separable:

Try though we may to split light into fundamental atomic pieces, it remains whole to the end. Our very notion of what it means to be elementary is challenged. Until now we have equated smallest with most fundamental. Perhaps for light, at least, the most fundamental feature is not to be found in smallness, but rather in wholeness, its incorrigible capacity to be one and many, particle and wave, a single thing with the universe inside.[44]

Whereas Zajonc relies primarily on scientific evidence to make his point regarding light's wholeness or togetherness, Merleau-Ponty eschews theory while seeking to "emulate the unreflective life of consciousness."[45] In that life, light is fully coincidental with our experience of light, and there is a sense in which light slips or timelessly traverses whatever interval separates observer and object. Although object and observer are opposite ends of a relation, each reaches into the other in virtue of the relation that conjoins and compounds them. Thus they may be said to touch and modify one another even though, from a Newtonian perspective, they occupy different locations.

Earlier I suggested that this relationality, or something like it, shows up in the scientific determination of nonlocality or quantum entangled particles. Remarkably, Merleau-Ponty, upon interrogating pre-theoretical experience, broached a similar conclusion regarding macroworld events: the observer-object interval that prevails at the reflective or theoretical level stands perpetually revoked—never effected—at the pre-reflective level.

11

EXISTENTIAL LIGHT

In our rhythm of earthly life we tire of light.

T. S. Eliot, "Choruses from 'The Rock,'" in T. S. Eliot,
The Complete Poems and Plays: 1909–1950

Although Husserl sought to return philosophy "to the things them-selves," to a world untouched by undue philosophical complication, some argue that he failed to accomplish this. As he continued to wrestle with fundamental issues, he seems to have shifted back to the notion of mind surveying the world from a privileged perch. What he came to call the "transcendental ego" may be understood as mind transcending the realm in which it was once fully situated. Martin Heidegger, one of his students, was careful not to repeat the mistake, if indeed it was one, while forging ahead with a phenomeno-logical analysis of his own. Let me rehearse Heidegger's contribution as it relates to the issue at hand. I also want to recount two mythic or religious variations on the Heideggerian motif of being as light-like expanse. Although Heidegger might not endorse these variations, they help us see in yet another way that light is too elemental for ordinary comprehension. While wave-particle duality is perhaps the best known expression of light's ability to elude precise definition, there are others. In this book we have focused on the speed of light, a finite value that nevertheless cannot be reduced to unambiguous, finite understanding. Now we turn to a more existential expression of light's bivalence or ambiguity: its presence consists of its absence, an absence that clears space for reality or creation.

Heidegger's *Lichtung*

Heidegger put great stress on *being-in-the-world*, an expression roughly equivalent to *Lebenswelt*. To be is to be situated in the world, and our situatedness—our being—is coincidental with our perceptually expansive awareness of the world. What is more, because our being precedes and fosters intellectual analysis, it cannot be captured thereby. By talking about what occurs before philosophical or scientific reflection gets underway, Heidegger showed that most of the time we live in a world very different from the one given us by science. As I go throughout my day, things smoothly articulate into a coherent whole—at least most of the time. Occasionally something will malfunction or break down, and then that object—a stapler, say—gets singled out for special attention. This episode, or the recurrence of such episodes, persuades me that the stapler is a self-standing object in the world, and I then generalize that other objects are as well. I will probably even class myself among these objects, and so the world emerges as an aggregation of disjointed objects, myself included. What is forgotten is that my primordial being-in-the-world furnishes the ground for this new attitude toward the world and that the objects I now take as primary phenomena "stand already in the light of Being."[1] That is, before the world fragmented into separate things and processes, it was a coherent whole. Moreover, once the crisis is over, once the stapler is fixed and I resume my pre-reflective posture toward the world, it goes back to being a coherent whole. As I bicycle home from work, the world presents itself as an intricate harmony of sight, sound, texture, and smell. Not that I have a mystical vision; this is just ordinary reality, and I am expansively integrated into it.

In developing his analysis, Heidegger hugged the shore of everyday experience. He realized that even philosophers and scientists, notwithstanding their "theories of everything," live off the top of experience that cannot be made theoretically explicit. Hence his concern with ways of having and knowing in advance of intellectual having and knowing. We come to understand the world through a theoretical lens precisely because we are cognate with the world. We are at home in the world long before we adopt a theoretical posture toward it, and even this reaches back into pre-theoretical experience like a wave reaches back into the ocean. What is more, the very thing that enables science—being-in-the-world from the start—is hard to grasp owing to its deep familiarity. When things are smoothly articulating, they get iterated out of sight; they become transparent. Heidegger's favorite example of this is a man using a hammer. When all is going well, the man scarcely thinks of the hammer. It is there, of course, but it has been drawn into a large scheme of meaning—perhaps the man is building a cabinet. As long as the hammer functions properly, the man will not take explicit notice of it. He will stare past it in his quest to finish the cabinet.

Heidegger proposed that, owing to our being-in-the-world, our meaningful involvement in things, a similar transparency takes shape, albeit globally. In various ways being gets covered over or stared past; it gets re-veiled even as it gets revealed. Perhaps, as just suggested, being gets assimilated to things like staplers so that a thinghood ontology emerges. So instead of being an expanse within which things show up, one's existence is taken as just another thing. "Why is existence weird?" asks Michael Zimmerman while explaining Heidegger. "Because humans are not things, but the clearing in which things appear. Although we are not fixed things, we define ourselves as if we were simply a complex version of the things we encounter in the world: rational animals."[2] So defined, we misconstrue ourselves, believing that we are incidental observers in the world. In fact we are a "centered" and "absorbed openness" in the world, and that implies that our being cannot be subtracted out of the world.[3] Each presupposes and leans into the other.

Heidegger gave being an explicit optical and revelatory characterization. It is *Lichtung* (lighting, clearing) and *Offenbarung* (opening, disclosure, revelation). And because being is all-involving and already there from the start, we do not stand on the sidelines and passively view it. Each human is what Heidegger called *Dasein*, a being absorbed in the world despite his or her local particularity. *Da* denotes hereness or thereness, inescapable locality at the center of the world: in moving about I never step away from myself. *Sein* means being, which when coupled with *Da* connotes absorption beyond one's local centeredness. Although centrally located, I spontaneously spill over that centrality through some sort of un-self-containment. At this moment I am thrown into a wide embrace of books, furniture, wall objects, windows, trees, streets, mountains, and clouds. This is the centered openness or *Dasein* that is me.

So described, *Dasein* recapitulates some of the paradoxes of light. Inescapable locality lies at the heart of special relativity, and it is secured by Einstein's second postulate, whereby space and time are subordinated to the constancy of the speed of light. Without that subordination, observers may assume that all local events occur on the same universal page (that of absolute space and time) coincidental with an otherworldly or God's-eye view of nature. But once light speed constancy takes effect, events can be mapped only on local pages corresponding with different human experiences within the cosmos. Situated in the world, we cannot transcend it, and so it follows that certain constancies might arise at the interstices of our being in the world. Light speed constancy seems to be one of these, as does Heidegger's *Dasein*: not something we can get intellectual leverage on, *Dasein* cannot be de-centered from our experience. Something light-like constitutes the constancy of being, the constancy that *is* being, said Heidegger, for being means "appearing" or "standing-in-the-light."[4]

Furthermore, because of *Dasein*'s primordiality it is not something with a bottom or discrete boundaries—it is open. But, then again, since it is you and me, it would not be something apart from us, something we can bump up against in an objective, knowing way. If light is "the 'letting-appear' that does not itself appear,"[5] being is the letting-be that does not explicitly register as anything in and of itself. Being self-reflexive, it cannot hold still for itself, and this is why Heidegger described it as unsettled, groundless, and indeterminate. In a way it is nothing at all, but nevertheless something: a clearing that doubles both as being and as an opening or emptiness whereby being emerges or becomes. It is the "emerging sway" that "includes both 'becoming' as well as 'Being' in the narrower sense of fixed continuity" or constancy.[6] If it were not there from the start, there would be no basis for existence, but if it were also *not* there from the start, *ongoing* existence would be ruled out. Being would not be able to complete itself ad infinitum. Thus its indeterminate or groundless nature, its "falling away" from what it is. In one passage Heidegger describes being as "an oblivion that itself falls into oblivion."[7]

As noted, Heidegger's word for clearing or opening is *Lichtung*, which may readily be translated as light or lighting. But rather than yielding to the popular notion that light is just another thing—something that comes along with or even after myriad other things—Heidegger chose to underscore *Lichtung*'s primordiality. It is dawning—a dawning or glory that "constantly persists" as a never-dying backdrop and foundation for all else.[8] "The first time we see *light*," wrote William James, "we *are* it rather than see it."[9] To see light is to be initiated into the open set of things—so much so that we are that set or revelatory clearing, the dawning that precedes the sifting, sorting process whereby *Lichtung* is eventually sunk beneath an assemblage of things or entities, each of which (as the word *entity* connotes) appears fully self-existent. What thus gets covered over or closed off is the opening whereby things manifest themselves. *Lichtung* is inclusive of those things but also more elemental, and owing to its open, indeterminate nature, always in danger of being overlooked and forgotten.

The One and the Many of Light

"The central paradox of Heidegger's philosophy," writes Frederick Olafson, "stems from the fact that he wants to say that Dasein is the clearing and also that being is the clearing. If Dasein is inherently plural and being is just as inherently singular and unique, it is not apparent how both these assertions can be true."[10] As individual Dasein, each person is an "absorbed openness" whose unique centrality gives him or her singular command of the field of being. But there are many Dasein and only one being. How, Olafson wants to know, can every Dasein, situated differently, be the pivot around which being uniquely swings?

Responding to this question, Theodore Schatzki points out that Dasein, while identified with individual and numerically distinct human beings, also refers to (in Heidegger's phrase) "This entity which each of us is himself."[11] That is, Dasein is a way of being which humans, in their plurality, uniquely share. Each of us experiences a certain mineness which spirals in toward particularity and individuation. At the same time, however, that mineness is universal among thinking human beings, and this keeps Dasein from wholly fragmenting into multiple parts, each of which, it would seem, would be a world unto itself. Thus, individuals are "instances of Dasein,"[12] iterations of a unitary way of being. What is more, Schatzki continues, "certain commonalities" among individual Dasein tend to bring those Dasein toward singularity. Intrinsic to Dasein are aspects of sociality and referentiality (*Mitsein* and *Das Man*), whereby one as distinct individual bridges into one as single, mutually constituted being. Thus, says Schatzki, "[individual] clearings are largely the same" because they are, in large measure, identically structured.[13]

This all makes sense, but Heidegger's identification of Dasein and being with clearing or *Lichtung* invites further analysis. We have already noted that photons, once imagined as individual parts of light, are not always denumerable and separable. Rather they come off as seamless and expansive, more like the light of everyday experience than discrete parts of light confined to particular locations and well-understood modes of action. This, however, is merely a descriptive parallelism. We still do not know how or why each Dasein is a unique pivot around which the world swings. We only know that something similar happens at the level of light and that Dasein is light-like.

Light speed constancy, however, affords some understanding. As noted earlier, if you and I are at different locations, a spark at the center of your light-sphere cannot be at the center of mine. Once I see that spark, however, its light will radiate away from me omnidirectionally at the speed of light, thus situating me at the center of things. Hence each perspective, each point of reference, is, when occupied, a central point around which the world pivots unerringly. And with no perspective that trumps all others, each is prime when occupied. This primeness or centrality, I submit, is not unlike Dasein's way of being, its "centered openness"[14] that coincides with its sense of mineness or individuality. And yet that sense is not decisive, according to Heidegger, because it is a unitary or shared way of being. Thus individual, denumerable Dasein cannot be cleanly distinguished from each other, just as individual photons, seemingly self-contained and countable, cannot, in all circumstances, be said to be fully distinct from one another.

We may go further. Owing to light speed constancy, light comes off as a centering principle, thereby turning every occupied spacetime location into a focal point—seemingly *the* focal point—of all reality. But this is more than an abstract idea; it harks back to the remarkable fact—at least it seems remarkable at first—that our tiny, point-like eyes can take in vast swaths of visual detail at a single glance. Is this merely a function of individual photons, each of which conveys a piecemeal image, converging on the retinas? Or might each photon indivisibly (nonlocally) share itself with other photons so as to offer observers a fully centered and integrated vision of reality?

Light's universal capacity to center observers, I am suggesting, unfolds from its ability to inform all of its parts with the fullness of its reality; to be, in Arthur Zajonc's phrase, "a single thing with the universe inside."[15] As Galileo proposed regarding sunlight, "the sum of all its business" appears to be the ripening of a single grape, notwithstanding countless other grapes.[16] More prosaically, one might say that the sum of all light's business is to infuse every occupied spacetime location—every Dasein—with centralized cosmic awareness.

In making this claim I am keying off others. After asking how "immense stretches of space and time" may be "contained in the movements of light in the tiny space encompassed by the eye," David Bohm suggested that the "*total order* [of reality] is contained, in some *implicit* sense, in each region of space and time."[17] Henri Bortoft, inspired more by Johann Wolfgang von Goethe than by Bohm, makes the same argument by referring to the experience of looking up into the starry heavens:

> We see this nighttime world by means of light "carrying" the
> stars to us, which means that this vast expanse of sky must all
> be present in the light that passes through the small hole of
> the pupil into the eye. Furthermore, other observers can see
> the same expanse of night sky. Hence, we can say that the stars
> seen in the heavens are all present in the light that is at any
> eye-point. The totality is contained in each small region of
> space.[18]

Bortoft concludes by saying that "[i]f we set off in imagination to
find out what it would be like to be light, we come to a condition
in which here is everywhere and everywhere is here. The nighttime
sky is a 'space' that is one whole, with the quality of a point and yet
including all within itself."[19]

Bortoft finds part and whole interchanging within a light-based
visual economy. Heidegger finds something similar within the clearing or *Lichtung*. When its implications for observers are drawn out,
light speed constancy allows one to grasp, at least in a provisional
way, how every Dasein can be distinct from every other and yet a
centered openness around which the world uniquely pivots.

Light as Nothingness

The somethingness of nothingness is an old theme that finds many
illustrations, some of which evoke light's uncanny behavior. We have
already noted that light shows up only in conjunction with material bodies, that it is the one thing that can be added to a vacuum
without ruining the vacuum, and that it drops out of sight to give us
sight. In a word, light *per se* is eventless or invisible. Yet without it
we would see nothing. It enables vision while remaining, in and of
itself, visual nothingness.

But this should not surprise us, for we find the same principle
playing itself out in other settings: nothingness at the eye of the perceptual or cognitive storm. John Locke recognized the problematic
character of the mind's attempt to grasp its own nature by reflecting that although the eye is made for seeing, it cannot see itself, at
least not in an immediate or direct way.[20] This logical problem of
self-reflection prompts the following question: if the eye were not

self-blind, would it be able to see beyond itself? Perhaps not, for it would seem that the eye's immediate experience of itself would be a world unto itself. Aristotle proposed as much with regard to the mind (*nous*): "For if [*nous*] shows its own identity, it hinders or obstructs what is other than it; hence it can have no nature but that of capacity. What is called *nous* of the soul, then . . . is not anything until it knows."[21] Accordingly, the mind "*is not*, has no content until it knows [something other than itself]."[22] If one substitutes being for mind, this outlook lines up with Heidegger's view that without being-in-the-world, being would never emerge. Not only that, but it suggests that an elemental self-innocence enables the mind's work of knowing the world. Not being a self-enclosed or self-knowing system, the mind is open to the world. Following Heidegger, I am proposing that this openness is light or *Lichtung*. It is a kind of self-nothingness or self-innocence, even self-forgetting, that keeps the world open, uncontained, and indeterminate.

Heidegger argued that the ancient Greeks were more mindful of this openness or *Lichtung* than we are. But that was before the world became a victim of human surveillance and human enlightenment, before something happened to "progressively swing the brilliance of Being toward human vision" with its tendency to sift and separate.[23] Brilliance here denotes not just optical radiance but also the way visible things coherently hang together despite intervening visual nothingness. More than scattered images of reflected light, the visible world is light-articulated expanse. Light coordinates visible objects, brings them onto the same visual page, but not in a self-announcing way.

Even before Heidegger or the ancient Greeks, people envisioned light as the opening of new possibility or, to use the mythic idiom I will now employ, as the clearing of space for creation. In ancient Egypt, day and night consisted of being drawn into and expelled from the vision of Ra, the sun-god. In Ra's words, "I am he who openeth his eyes, becometh the light; shutteth his eyes, becometh the dark."[24] The phenomena of nature, which became startlingly diverse and multi-colored at sunrise, were part of Ra's consciousness. The world therefore, despite its immense variegation, was purportedly kept intact by the light-mediated consciousness that framed and suffused it.

Roughly coincidental with the light of Ra was the light of Shu. When Earth and Sky (Geb and Nut) were locked in sexual embrace, it was Shu who got between them to open up an expanse of space and light within which the created order could come into being. Often characterized as wind because of wind's capacity to rearrange and expand, Shu was also regarded as the light of creation—that which brings about not only the activation of life processes but more fundamentally the clearing or opening that grants life an opportunity to emerge and stretch forth. In this latter sense, Shu's light recalls *Offenbarung*, the German word for revelation that literally means opening. Amid the gods, Shu opened up or lit up a space for humans and thereby brought them into the divine consciousness that filled the cosmos.

In time, however, and in a manner approximating Heidegger's analysis, the light of Shu and Ra was gradually overtaken by lighted things so that individual consciousness emerged to wean itself free of divine consciousness. The very first cosmogonies describe the world's creation in physical terms: water, wind, spit, semen, or other physical substances were hailed as the seed principles of creation. When, however, pharaohs of the First Dynasty moved the center of the world (the religious capital) from Heliopolis to Memphis, they rationalized the change by introducing Ptah, a creator-god who thought and uttered the world into existence. Further, Ptah was given primacy over other creator gods; that is, his act of intellectual conception preceded and inspired their acts of physical conception.

The import of this development, writes John A. Wilson, lies in the "attempt to relate creation to the processes of thought and speech rather than mere physical activity."[25] In brief, this is "an approach to the Logos Doctrine,"[26] the saying of the creative Word that initiates the physical creation. Though word is distinct from world, creation passes from one to the other along a single trajectory: the luminous cosmos points back to a luminous thought. Human creativity subsequently came to mean riding in the intellectual wake of the divine creation, drafting on the creation to partake of its revelatory splendor.[27] Instinct with re-experiencing the creation, we tag along, mimetically composing song and art.

This was the aboriginal understanding: poetry and song was an echo of logos, and so the poet caught his song "on the rebound."[28]

The ability to think creatively had everything to do with the splendor of creation. At some point, however, human creativity began to veer off on its own trajectory, disavowing its mimetic dependence on the world. It was as if the moon claimed to be self-luminous. Creative thought, once regarded as a mode of revelation about the world, asserted its independence from the world. Bright ideas, we now believe, just issue up from human genius; they are not caught on the cosmic rebound or given to us by the bright world.

In George Steiner's phrase, "the contract or covenant between word and world has been broken" so that "[t]here is in words and sentences no pre-established affinity with objects, no mystery of consonance with the world."[29] The result is the modern sense that, when it comes to language, anything goes. Like a deck of cards, words and ideas can be shuffled at will, for the world, though physically present, no longer has an originary Word or Event (*Ereignis* in Heidegger's vocabulary, a word connoting *Lichtung*)[30] which structures and keeps it on track, and that is why human creativity no longer feels itself nourished by or even linked to the moment of creation.

What, after all, is "the moment of creation" for moderns? The phrase no longer means much. Michael Welker argues that modern cosmologists dimly grasp the question of creation, at least as that question was understood in antiquity. They think of creation as an "initial ignition," not as an integrative event whose force is still fully felt in the way the cosmos is organized and held intact.[31] The result is a sterile, imaginatively stripped-down creation account like the big bang: a remote, uncontextualized flash of light having only accidental connection with human life. All Beginning and no End or purpose.

By raising thought and speech above physical elements like wind and water, the priests of Ptah launched humankind on the heady adventure of rethinking the thoughts of God at the creation. As just suggested, however, that adventure acquired a life of its own, and Egyptian mythology hints at the first stirrings of intellectual self-determination. The baffling story of the Eye, against whom humankind rebelled, is, according to Rundle Clark, "the key" to the Egyptian religion.[32] This was not a modern passive eye, but a Divine Eye radiating a fiery, almost blinding intelligence. Detached from the face of the High God in order to seek out two lesser deities who

Figure 11.1. Two views of the uraeus.

were lost in the primeval abyss, it returned to find itself supplanted by another eye—the sun-god Ra. Momentarily swept up in the joy of seeing its face again and bringing the lost gods back home, the Eye wept tears that became the primal parents of the human race. It then became angry at its displacement and took the form of an enraged cobra. Ra, hoping to appease the Eye, wrapped the cobra around himself. Egyptian pharaohs, tracing their authority back to Ra, had their crowns fashioned in like manner: the bright eye of heaven protected and legitimized by the visible memory of the Original Eye. This is the uraeus symbol depicted in figure 11.1, the royal emblem.

Clark calls the Eye "the striking-power of the High God" and interprets the story as humankind's attempt to strike back.[33] In an era when the universe was brimming with divine intelligence, humans sought to tame that intelligence and democratize the relationship between themselves and the gods. The intense and even overpowering gaze of heaven had to be moderated; otherwise there would be no space for human initiative. According to Clark, "the pacification of the Eye is the establishment of monarchy."[34] With pharaohs as gods or agents thereof, humans could encroach upon divinity.

The story of the Eye prefigures the Greek tale of Prometheus. Not content with his human lot, Prometheus stole fire from heaven and subsequently was punished by Zeus for his presumption. That this episode involves more than the invention of physical fire is indicated by Prometheus's name, which means "forethought." Unlike his brother Epimetheus, whose name means "afterthought," Prometheus is creative in the modern sense: his thought has a forward momentum that propels it beyond the bounds of divine decree. With an eye toward his own piece of the divine action, he schemes, plans, and analyzes, and thus abandons the more contemplative "afterthought" mode of being whereby insights are caught on the cosmic rebound. In a large sense, Prometheus learns to think for himself—learns to kindle his own intellectual fire rather than bask in the fire-illuminated expanse of the gods.

We customarily celebrate this tale as a heroic fable about humankind's coming of age in a universe that is mistakenly believed to be ruled by capricious gods. This is only half the story, and a somewhat distorted half at that. Possibly the Egyptians, with their lively apprehension of light, knew the losses entailed by the gains. The biggest loss was the eclipse of divine being as human beings (pharaohs) began to stand forth as objects of veneration. Implicit in this loss is that of light, for few people now appreciate the momentous implication of the creation story of Shu: though light is divine, it is light to the human touch. Amid the gods, light graciously clears space for human initiative. Thus the creation refers to an ongoing process whereby divinity accommodates humanity, even as humanity becomes steeped in its illusion of self-luminosity. This graciousness, which is central to the paradoxical essence of light, has been obscured by things graciously made visible by light.

Kenosis

All cosmologies must wrestle with the question of nothingness, for often nothingness seems to be something. The Greek atomists were perhaps the first to bring the problematic nature of nothingness into philosophical focus by paradoxically insisting that the absence of reality is an indispensable feature of reality. Their initial inclination was to argue that reality consisted merely of atoms, tiny bits of physical matter. But the flux of nature pointed to more than just atoms. If reality were simply atoms, motion would be impossible, for free space would not exist. It became necessary, therefore, to introduce a non-material reality—the void—as a kind of playfield for atoms.

In some religions, physical reality's paradoxical dependency on nothingness can be traced back to an act of divine self-sacrifice. This motif also shows up in the atomist world-view, though much subdued and lacking explicit moral import: if atoms are to be active and penetrative, something has to be passive and receptive. The void must exist, but only in a way that allows material reality to stand forth as the cynosure of all eyes.

This seems to be the dialectic that informs light. Thanks to light, we see the world, but only because we do not see light. It is the hole or void in our sensible experience that enables vision, an emptiness that receives the cosmic spectacle, a self-vacating presence filled by the Other, a vacuum that allows creation to rush in. Or, to switch images one last time, like the number zero in a string of integers, light offers no value of its own, but by holding an "empty" spot in the order of things, it decides the value of all else.

For those inclined to hear the resonance, these formulas chime into certain creation stories that posit divine self-negation or self-withdrawal as a precondition of cosmic genesis. This may sound dismal, but because of the grace or compassion that attends that self-withdrawal or sacrifice, the mystery of creation has a positive pole as well as a negative one. Widely associated with creation, light as we have come to understand it seems to embody that duality.

Here an anecdote may prove helpful. A teacher once offered a silver dollar to any of his students who could tell him where God was. After a moment of silence, a young girl raised her hand and said: "I will give two dollars to anyone who can tell me where God is not." Some might propose that light is the place God is not. He is not there because he voluntarily went into exile at the creation, and light's invisibility and empirical intractability instantiate his exile or absence from the universe. But this is only half the story, because we, as participants in the cosmos, are beneficiaries of God's loss. That loss is not just our gain, but the very fount of our existence. In it (to borrow from St. Paul), "we live, move, and have our being."[35] There is, then, something profoundly positive about God's absence from the universe. It turns out to be an expression of grace or sacrificial caring, and so God's goodness hangs in the air, as it were, despite his absence. The result is light—a very light divine presence that allows creation to spring into existence.

To grasp this point in more detail, we turn to Joel Primack (a cosmologist) and Nancy Abrams (a musician, novelist, and lawyer married to Primack). In the hope of elaborating a wider understanding of our cosmic origins, Primack and Abrams have noted parallels between modern cosmology and a particular kabbalistic creation narrative. Their aim is not to suggest that kabbalistic concepts prepared the way for modern science, but rather to offer "a spiritual analogue of theoretical physics" that can resonate mythically in modern culture.[36]

One of the parallels noted by Primack and Abrams is that of some kind of exit or fall from an eternal state of being. Among physicists who subscribe to the idea, this hypothetical state is called "eternal inflation." It is a realm of accelerating expansion and superabundance—"an eternal blizzard of universes."[37] In stark contrast to our present experience, eternal inflation marks a domain where "sparkpoints" or seeds of larger possibility multiply and grow in limitless profusion. A weak analogue to this riot of endless creativity might be a child's imagination, which is open to all sorts of possibilities, however improbable. But just as a child grows up and eventually realizes that the world cannot accommodate all of one's dreams, so some "dreams" or sparkpoints go unsustained in the eternal realm.

Our universe, suggest Primack and Abrams, is one such sparkpoint. Failing to catch any of the inflationary waves of eternity, it spiraled down toward oblivion. Finally breaking into a different order or disorder of things, it momentarily flared up to produce the more conventional inflation that some believe characterizes the first moments of the big bang. Then the flare subsided and star systems began to form.

By this account the universe is a spark or seed of light cut off from a bonfire of staggering creativity. And having fallen into an alien economy as "an almost vanishingly small capsule of eternal creativity,"[38] it has set free some of that creativity in the stars, planets, and life forms we see about us. But for all its material grandeur, the present universe fails to convey the terrible reality of eternal inflation with its self-multiplying infinities.

The word *eternal*, of course, has a spiritual ring to it, and so for some the scientific notion of eternal inflation invites religious contemplation. Primack and Abrams act on this invitation by observing that around 1570 Isaac Luria extended kabbalistic thought by "teaching that in the beginning, God began to withdraw into self-exile in order to make space for the universe."[39] This act of divine self-contraction, known as *tzimtzum* in Kabbalah, amounted to God's exit from eternity. Thereby a void was left for the genesis and expansion of the cosmos, which God quickens by remaining veiled or hidden up in the void. His supernatural glory weakened, God limps, but creation abounds.

This then is a response to the theological question of how the finite, temporal cosmos fell, as it were, from the lap of eternity: God voluntarily retreated into a background of anonymity so that the creative order might stretch forth and flourish. In sharing this response, Primack and Abrams offer tentative support for the scientific hypothesis of eternal inflation. Their goal, however, is not so much to affirm a particular scientific or religious outlook as to provide material for a "functional cosmology" that is both scientifically and spiritually relevant.[40]

Taking this argument a little further, let me point out that light seems to embody the gracious, self-denying aspect of Luria's creator God. Just as that God backed off to let other, lesser realities emerge, so light goes into retreat that material objects may emerge. Were it not for light's hiddenness or invisibility, they would never be seen. In this regard, light is very good evidence of a very good Creator God. But here I must use "very good" strictly in the sense of benevolence, which makes light, with its invisibility, poor empirical evidence of God's existence. A dictatorial God would probably overpower us with his presence; there would be no question of his existence. But there would also be no freedom to ask such questions and to aspire to higher experiences and understandings. While we would have perfect empirical evidence of God's existence, we would have no opportunity to stretch forth and flourish. As it is, our empirical evidence for God's existence is far from perfect, but that is because the cosmos came about as God graciously *withdrew* from the scene of divine action, leaving only a spark or trace of his erstwhile presence to enkindle Creation.

God's creative goodness, in other words, militates against proof of his existence by empirical means, for God is not a heavy-handed Creator. He is light-handed; that is, his presence is light, and that lightness constitutes an opening in the divine plenum for the materialization of a non-divine cosmos. What this implies, among other things, is that creation ex nihilo—creation out of nothing—need not exclude creation out of something. God's self-created emptiness or nothingness represents a loss for God, and that is something. That loss, moreover, opens the creation drama.

This characterization of light, a vanishing act of divine munificence, coincides with the pre-modern sense that light is superabundantly or inexhaustibly real, notwithstanding—or perhaps owing to—its uncurtailed blankness. "Light remains what it is," says Hans Blumenberg, "while letting the infinite participate in it; it is consumption without loss. Light produces space, distance, orientation, calm contemplation; it is the gift that makes no demands, the illumination capable of conquering without force."[41]

Similar appreciation of light-like nothingness shows up in Christianity. In his epistle to the Philippians Paul wrote that Jesus Christ, though in "the form of God," "made himself of no reputation, and took upon him the form of a servant, and was made in the likeness of men." This expression coincides with many others in the New Testament, but in the original Greek the phrase "made himself of no reputation" may be literally translated as "emptied himself." Thus the idea of *kenosis* (from the Greek cognate meaning emptying), which has provoked much discussion among scholars intent on defining Christ's nature. Commenting on the passage in question, John Calvin wrote: "In order to exhort us to submission by His example, he [Paul] shows, that when as God he might have displayed to the world the brightness of His glory, he gave up His right, and voluntarily emptied Himself; that he assumed the form of a servant, and, contented with that humble condition, suffered His divinity to be concealed under a veil of flesh."[42]

Here *kenosis* is rendered two ways—as self-emptiness and as self-concealment—but each is concerned with the voluntary dimming of God's glory. When describing the opening or clearing of being, Heidegger spoke of a "self-withdrawing" whereby being is eclipsed and concealed by individual entities.[43] Equating being and truth, he further proposed that truth is nothing less than unconcealment: disclosure of that which is so primordial as to always be at risk of being forgotten or concealed. Thus, as suggested earlier by the cognate terms *reveal* and *re-veil*, truth goes both ways at once.

One sees this double movement in early-twentieth-century physics. In 1900 Lord Kelvin, a prominent British physicist, stated that just "two problems" obscured the "beauty and clearness of theory."[44] One problem was Michelson and Morley's failure to detect an ether wind altering the speed of light. The other concerned blackbody radiation, whose study led to quantum theory. Both problems were light-related, but when Einstein and others cleared them up, light itself became unclear or mysterious. Its speed became constant with respect to all observers and its nature no longer amenable to exclusive characterization as wave or particle. The two problems were erased, but light itself could not be clearly exposed; indeed, it seemed to undergo concealment. It remains in the erasures or clearings, but not in a self-showing way.

Summing Up

The clearing marks our being-in-the-world and consequently that aspect of our nature that cannot be lifted out of experience for clear— that is, objective—viewing. It is what drops out of sight in the seeing process, the slippage that affords vision. This outlook, of course, challenges the longstanding assumption that we can back away from nature, even our own nature, to comprehend it fully. Heidegger felt otherwise. As beings in the world, he reasoned, we are unable to cleanly separate ourselves from the world. We are, in other words, blurred or organically fused into the world, and this in turn implies blurred or indeterminate understandings of our situation.

I have argued that light speed constancy is an indeterminate understanding arising from our inescapable involvement in the world. By the time we get around to measuring the speed of light, we have prior (pre-measurement) familiarity with light and are therefore already participatory in its motion—hence we do not move relative to it. If space and time held absolute sway, as Newton believed, perhaps we could separate ourselves from the material world. But Einstein insisted on the absoluteness of the speed of light, arguing that this invariant value structures not just the spacetime cosmos but our experience thereof. To take stock of light speed constancy, then, is to glimpse the possibility that our powers of observation are cognate with the things we observe. As we move relative to light, we are perpetually reabsorbed into a relation older and more fundamental than relative motion. Light speed constancy is a constant of nature inclusive of our nature.

It is also a reminder that space and time do not exhaust the world's reality. On the scene before space and time, light conserves its ontological priority by regulating the motion and properties of material bodies. At a more elemental level, that of pre-reflective visual experience, it reenacts the dimensionless economy of the big bang singularity by giving us immediate access to the world. Intervals we once took as absolute and inviolable are found, on closer examination, to be ineffectual. Light is bound up in truths that transcend space and time, and as light-infused beings, so are we.

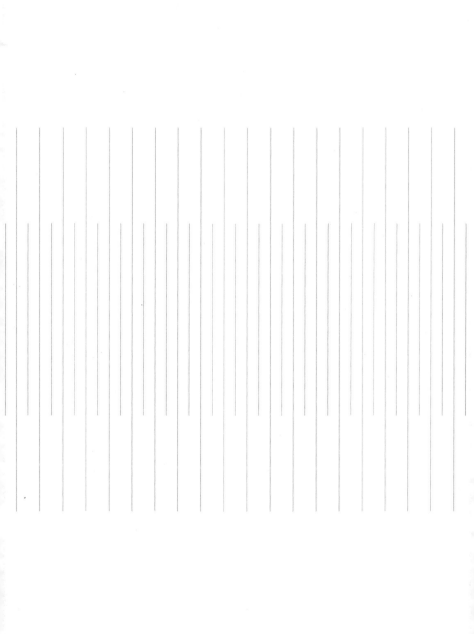

NOTES

Introduction

1. Let me offer two explanatory notes at the outset. First, in this treatment the word *light* is shorthand for all electromagnetic radiation. Second, when Albert Einstein developed his special theory of relativity, he postulated that the speed of light is constant in all inertial (unaccelerated) reference frames. This constancy subsequently carried over into his general theory of relativity, which concerns all reference frames, inertial and non-inertial. In either theory the speed "falls out" of the geometry of spacetime and is therefore an incidental property of light. That is, it is not as if light moves against an indifferent background of space and time; rather the speed of light already structures space and time before it shows up as a value exclusively associated with light. For ease of expression, I will generally say "all reference frames" when speaking of light speed constancy.

2. N. David Mermin, *It's About Time: Understanding Einstein's Relativity* (Princeton, N.J.: Princeton University Press, 2005), p. 25. Emphasis in the original.

3. Hans Reichenbach, *From Copernicus to Einstein*, trans. Ralph B. Winn (New York: Philosophical Library, 1942), pp. 67–68.

4. Jacques Derrida, "Faith and Religion: The Two Sources of Religion at the Limits of Reason Alone," in *Religion*, ed. Jacques Derrida and Gianni Vattimo, trans. Samuel Weber (Stanford, Calif.: Stanford University Press, 1998), p. 16. Derrida parenthetically equates "revealability" to the German word *Offenbarkeit* and "revelation" to *Offenbarung*. For readability, I have deleted these insertions.

5. Hans Blumenberg, "Light as a Metaphor for Truth," in *Modernity and the Hegemony of Vision*, ed. David Michael Levin (Berkeley and Los Angeles: University of California Press, 1993), p. 31.

6. In modern cosmology a discontinuity is said to exist between the big bang singularity (where spacetime, having infinite curvature, does not

comply with the laws of physics) and the relatively flat spacetime universe that follows in its wake. As Stephen Hawking puts it: "At a singularity, all the laws of physics would have broken down. This means that the state of the universe, after the Big Bang, will not depend on anything that may have happened before, because the deterministic laws that govern the universe will break down in the Big Bang. The universe will evolve from the Big Bang, completely independently of what it was like before." Stephen W. Hawking, "The Beginning of Time," http://www.hawking.org.uk/lectures/bot.html, accessed 26 June 2005. I am suggesting that the discontinuity is not, as Hawking states, complete. It carries over into the universe as light, which, as we will see, is both continuous and discontinuous with the spacetime regime.

7. William James, *Varieties of Religious Experience* (New York: Vintage Books, 1987), p. 409.

8. Plato, *The Republic*, trans. Benjamin Jowett (New York: Vintage, 1991), p. 256. Translation modified: the pronoun "it" (referencing the sun) replaces "him" in Jowett's translation, which is mindful of Greek's gendered nouns.

9. Gerard Manley Hopkins, "God's Grandeur," in *The Major Poems*, ed. Walford Davies (London: J. M. Dent and Sons, 1979), p. 64.

10. John Archibald Wheeler, *A Journey into Gravity and Spacetime* (New York: Scientific American Library, 1990), p. 43.

11. Maurice Merleau-Ponty, "Eye and Mind," *The Primacy of Perception* (Evanston, Ill.: Northwestern University Press, 1964), p. 166.

12. Ibid., emphasis in the original.

13. William James, *Psychology* (New York: Henry Holt, 1910), p. 14. Emphasis in the original.

14. J. T. Fraser, *The Genesis and Evolution of Time: A Critique of Interpretation in Physics* (Amherst: University of Massachusetts Press, 1982), p. 39.

1. Space, Time, and Light Speed Constancy

1. Richard S. Westfall, *The Life of Isaac Newton* (Cambridge: Cambridge University Press, 1993), p. 168.

2. In his Scholium to *Philosophiae Naturalis Principia Mathematica*, Bk. 1 (1689), Newton writes: "I do not define time, space, place, and motion, as being well known to all. Only I must observe, that the common people conceive those quantities under no other notions but from the relation they bear to sensible objects. And thence arise certain prejudices, for the removing of which it will be convenient to distinguish them into absolute and relative, true and apparent, mathematical and common." He then offers this definition of time: "Absolute, true, and mathematical time, of itself, and from its own nature, flows equably without relation to anything external, and by another name is called duration: relative, apparent, and common time, is some sensible and external (whether accurate or unequable) measure of duration by the means of motion, which is commonly used instead

of true time; such as an hour, a day, a month, a year." The translation is by Andrew Motte (1729) as revised by Florian Cajori (Berkeley and Los Angeles: University of California Press, 1934).

3. Alexandre Koyré, *From the Closed World to the Infinite Universe* (Baltimore, Md.: Johns Hopkins Press, 1968), pp. 155–89.

4. Charles W. Misner, Kip S. Thorne, and John Archibald Wheeler, *Gravitation* (San Francisco: W. H. Freeman, 1973), p. 5.

5. Wheeler, *A Journey into Gravity and Spacetime*, p. 13.

6. Isaac Newton, *The Mathematical Principles of Natural Philosophy* (1729), Newton's Principles of Natural Philosophy, Dawsons of Pall Mall, 1968, Definition III. Opening page.

7. This is the Scholium to the definitions in *Philosophiae Naturalis Principia Mathematica*, Bk. 1 (1689).

8. Cited in Richard S. Westfall, *The Construction of Modern Science: Mechanisms and Mechanics* (New York: John Wiley and Sons, 1971), p. 158.

2. Special Relativity

1. Albert Einstein, "On the Electrodynamics of Moving Bodies," in Arthur Miller, *Albert Einstein's Special Theory of Relativity* (Reading, Mass.: Addison-Wesley, 1981), p. 401.

3. Horizonal Light

1. With time dilation the relativistic factor is put in the denominator, so that as v approaches c, observed time gets larger or lengthened out.

$$\text{time (observed)} = \frac{\text{time (proper)}}{\sqrt{1 - \frac{v^2}{c^2}}}$$

2. Arthur Zajonc, *Catching the Light: The Entwined History of Light and Mind* (New York: Oxford University Press, 1993), p. 260.

3. Wheeler, *A Journey into Gravity and Spacetime*, p. 43.

4. Sydney Perkowitz, *Empire of Light: A History of Discovery in Science and Art* (New York: Henry Holt, 1996), p. 76.

5. Bernard Haisch, "Brilliant Disguise: Light, Matter and the Zero-Point Field," *Science and Spirit* 10, no. 3 (1991): 30–31.

6. Stanley L. Jaki, *Is There a Universe?* (Liverpool, UK: Liverpool University Press, 1993), p. 19.

7. Peter Kosso, *Appearance and Reality: An Introduction to the Philosophy of Physics* (New York: Oxford University Press, 1998), p. 96. Emphasis in the original.

8. David Bohm, *Wholeness and the Implicate Order* (London: Ark Paperbacks, 1983), p. 123.

9. As Maurice Merleau-Ponty puts it: "The openness upon the world implies that the world be and remain a horizon, not because my vision would push the world back beyond itself, but because somehow he who

sees is of it and is in it." *The Visible and the Invisible*, trans. Alphonso Lingis (Evanston, Ill.: Northwestern University Press, 1968), p. 100.

10. Hugh Nibley, *The Message of the Joseph Smith Papyrus: An Egyptian Endowment* (Salt Lake City, Utah: Deseret Book, 1975), p. 80.

11. Stephen E. Palmer, *Vision Science: Photons to Phenomenology* (Cambridge, Mass.: MIT Press, 1999), p. 15.

12. Milič Čapek, *The Philosophical Impact of Contemporary Physics* (New York: D. Van Nostrand, 1961), p. 184.

13. Wheeler, *A Journey into Gravity and Spacetime*, p. 13.

14. Freeman Dyson, *Disturbing the Universe* (New York: Harper and Row, 1979), pp. 248–49.

15. Werner Heisenberg, *The Physicist's Conception of Nature* (New York: Harcourt, Brace, 1958), p. 24.

16. Harald Fritzsch, *An Equation That Changed the World: Newton, Einstein, and the Theory of Relativity*, trans. Karin Heusch (Chicago: University of Chicago Press, 1994), p. 118.

4. Experiential Light

1. Eduard Ruechardt, *Light: Visible and Invisible* (Ann Arbor: University of Michigan Press, 1958), p. 197.

2. John Schumacher, *Human Posture: The Nature of Inquiry* (Albany: SUNY Press, 1989), pp. 113–14.

3. Zajonc, *Catching the Light*, p. 260.

4. Hans Jonas, "The Nobility of Sight, " in *The Phenomenon of Life: Toward a Philosophical Biology* (Chicago: University of Chicago Press, 1982), p. 136.

5. Jacques Derrida, *Writing and Difference*, trans. Alan Bass (Chicago: University of Chicago Press, 1978), p. 92.

6. George Greenstein and Arthur Zajonc, *The Quantum Challenge: Modern Research on the Foundations of Quantum Mechanics* (Boston: Jones and Bartlett, 1997), p. 35.

7. Some of the argument that follows is taken from my "Light as a Solution to Puzzles about Light," *Journal for General Philosophy of Science* 33 (2002): 369–79.

8. The similarity between the two words is found also in their Latin roots—*videre* (to see) and *dividere* (to divide). This suggests an etymological link, but no lexical aid I consulted offered one, perhaps because the two meanings seem unrelated. But in this regard see David Bohm's attempt to create a set of verbal descriptions consonant with quantum reality. In that system, the verb "vidate" (which he traces back to *videre*) denotes "a spontaneous and unrestricted act of perception" amounting to "the apprehension of totality." When, however, the prefix "di" is appended to "vidate," the idea of division or separation emerges. Accordingly, "di-vidate" refers to "the spontaneous act of seeing things as separate." Bohm, *Wholeness and the Implicate Order*, pp. 36–37.

9. Blumenberg, "Light as a Metaphor for Truth," p. 31.

10. Recall John Schumacher's point that "nothing, not even light itself, can bring us news of its upcoming arrival." Therefore light is always present in the process of announcing distant bodies. *Human Posture*, pp. 113–14.

11. In the ideal case this would occur. Because, however, mirrors (and pearls) are not perfectly reflective and because air scatters light, we get only a finite series of images. We should recall also that, given the finite speed of light, under even ideal circumstances infinite time would be required to achieve an infinite series of images. These caveats, however, do not militate against the above argument. Even if the interpenetration of photons does not diverge to infinity, any interpenetration whatsoever would be consistent with their nonseparability across spacetime.

12. Zajonc, *Catching the Light*, p. 299.

5. Relational Light

1. T. S. Eliot, "The Dry Salvages," in *The Complete Poems and Plays: 1909–1950* (New York: Harcourt, Brace, and World, 1952), pp. 130–37.

2. G. S. Kirk, J. E. Raven, and M. Schofield, *The Presocratic Philosophers: A Critical History with a Selection of Texts*, 2nd ed. (Cambridge: Cambridge University Press, 1983), p. 198.

3. Ibid., pp. 202–203.

4. Einstein, "On the Electrodynamics of Moving Bodies," p. 401.

5. Dante Alighieri, *Paradiso*, Canto 33, lines 85–87, in *The Divine Comedy of Dante Alighieri, A Verse Translation: Paradiso*, trans. Allen Mandelbaum (Berkeley and Los Angeles: University of California Press, 1982), p. 294.

6. For Grosseteste, see his *On the Six Days of Creation: A Translation of the Hexaëmeron*, trans. C. F. J. Martin. (Oxford: Oxford University Press, 1996).

7. Daniel J. Boorstin, *The Creators: A History of Heroes of the Imagination* (New York: Vintage, 1992), p. 4.

8. Thorkild Jacobsen, "Mesopotamia: The Cosmos as a State," in Henri Frankfort et al., *The Intellectual Adventure of Ancient Man* (Chicago: University of Chicago Press, 1946), p. 138.

9. Ibid., p. 138.

10. *The Egyptian Book of the Dead: The Papyrus of Ani*, trans. E. A. Budge (New York: Dover, 1967), p. lxxviii.

11. Fraser, *The Genesis and Evolution of Time*, p. 39.

12. Ibid. See also J. T. Fraser, *Time: The Familiar Stranger* (Amherst: University of Massachusetts Press, 1987), pp. 222–42.

13. Wheeler, "Law without Law," p. 189.

14. Bondi, *Relativity and Common Sense*, p. 108.

15. Torrance, "The Theology of Light," p. 75.

16. Ibid. Torrance is not the only contemporary thinker to draw theological insight from modern physics regarding light. Iaian MacKenzie writes: "Theology is obliged to take heed of the fact that from James Clerk

Maxwell onwards, and past Einstein, all that has been and is unfolded in the scientific disciplines only enhances all theological endeavour in perceiving the propriety of light not only as a symbol of God, but as an 'earnest' (in the biblical sense) of God's constancy as he is *in se*, existing as the eternal and undivided Trinity, the Father and the Son in the bond of constant divine love, the Holy Spirit, and as he is *ad extra* in the expression of his Being towards creation in constant love." *The "Obscurism" of Light: A Theological Study into the Nature of Light* (Norwich, UK: Canterbury Press, 1966), p. 60. See Lawrence W. Fagg, *Electromagnetism and the Sacred: At the Frontier of Spirit and Matter* (New York: Continuum, 1999).

17. Torrance, "The Theology of Light," p. 76.

18. Acts 10:34, King James Version.

19. Matthew 5:45, King James Version.

20. Torrance, "The Theology of Light," p. 76.

21. Galileo Galilei, *Dialogue on the Great World Systems*, in the Salusbury Translation, revised, annotated, and with an introduction by Giorgio de Santillana (Chicago: University of Chicago Press, 1953), pp. 378–79.

22. Torrance, "The Theology of Light," p. 78.

23. Zajonc, *Catching the Light*, p. 260.

24. Compare William James's comment: "Our reflective mind abstracts diverse aspects in the muchness [of reality], as a man by looking through a tube may limit his attention to one part after another of a landscape. But abstraction is not insulation; and it no more breaks reality than the tube breaks the landscape." *Some Problems of Philosophy: A Beginning of an Introduction to Philosophy* (Lincoln: University of Nebraska Press, 1996), p. 199.

25. Wheeler, *A Journey into Gravity and Spacetime*, p. 43.

26. Torrance, "The Theology of Light," p. 87.

27. Ibid.

28. Ibid.

29. Ibid., p. 76.

6. Internal Relations

1. Bohm, *Wholeness and the Implicate Order*, p. 174.

2. Ibid., p. 133–34. For the ubiquity of patterns in nature and how they come about, see Philip Ball, *The Self-Made Tapestry: Pattern Formation in Nature* (Oxford: Oxford University Press, 1999).

3. Scott Clark, *Japan: A View from the Bath* (Honolulu: University of Hawaii Press, 1994), p. 7.

4. David Bohm, *Quantum Theory* (Englewood Cliffs, N.J.: Prentice Hall, 1951), p. 169.

5. Alfred North Whitehead, *Science and the Modern World* (New York: Free Press, 1967), p. 18.

6. Alfred North Whitehead, *Process and Reality*, corrected ed. by David Ray Griffin and Donald W. Sherburne (New York: Free Press, 1985), p. 39.

7. Ibid., pp. 39–40.

8. Ibid.

9. Alfred North Whitehead, *The Concept of Nature* (Mineola, N.Y.: Dover, 2004), p. 145.

10. Ibid.

11. Ibid.

12. Whitehead, *Science and the Modern World*, p. 84.

13. Ibid., p. 54.

14. Ibid., p. 51.

15. Ibid., p. 91.

16. Ibid., pp. 91–92.

17. Ibid., p. 92.

18. Ibid., p. 91.

19. William Wordsworth, "Lines Composed a Few Miles above Tintern Abbey," ll. 92–101, in *The Complete Poetical Works* (London: Macmillan and Co., 1888; Bartleby.com, 1999.) www.bartleby.com/145/ (16 Sep 2002).

20. Whitehead, *Science and the Modern World*, p. 17.

21. Ibid., p. 18.

22. See, for example, Shimon Malin, *Nature Loves to Hide: Quantum Physics and the Nature of Reality, a Western Perspective* (New York: Oxford University Press, 2001).

23. Robert Nadeau and Menas Kafatos, *The Non-local Universe: The New Physics and Matters of the Mind* (Oxford: Oxford University Press, 1999), pp. 196–97.

7. Light in a Vacuum

1. Plato, *The Republic*, pp. 246–53. Many others could be cited in support of Plato. Jonathan Powers, for example, writes: "When we see an object we see patches of colour, of light and shade. We do not see a luminescent stream flooding into our eyes. The 'light' we postulate to account for the way we see 'external objects' is not given in experience; it is inferred from it" (*Philosophy and the New Physics* [London: Methuen, 1982], p. 4).

2. Zajonc, *Catching the Light*, p. 260.

3. Henning Genz, *Nothingness: The Science of Empty Space*, trans. Karin Heusch (Reading, Mass.: Helix Books, 1994), p. 24.

4. Werner Heisenberg, *Physics and Philosophy* (New York: Prometheus Books, 1999), pp. 53, 70, 147–60. Elsewhere Heisenberg writes: "The concept that events are not determined in a peremptory manner, but that the possibility or 'tendency' for an event to take place has a kind of reality—a certain intermediate layer of reality, halfway between the massive reality of matter and the intellectual reality of the idea or the image—this concept plays a decisive role in Aristotle's philosophy. In modern quantum theory this concept takes on a new form; it is formulated quantitatively as probability and subjected to mathematically expressible laws of nature" (quoted in Max Jammer, *The Conceptual Development of Quantum Mechanics* [New York: McGraw-Hill, 1966], p. 286).

5. Dyson, *Disturbing the Universe*, pp. 248–49.

6. Genz, *Nothingness*, p. 1.

8. Ambient Light

1. Einstein, "On the Electrodynamics of Moving Bodies," p. 401.

2. This claim, previously noted, was first put forward by Plato in book 6 of *The Republic*. More recently P. W. Bridgman has written: "The most elementary examination of what light means in terms of direct experience shows that we never experience light itself, but our experience deals only with things lighted. This fundamental fact is never modified by the most complicated or refined physical experiments that have ever been devised" (*The Logic of Modern Physics* [New York: Macmillan, 1955], p. 151). Other similar statements from a broad range of thinkers could be offered, but I hope these will suffice to make the general point.

3. Addressing the apparent ability of photons and other subatomic particles to remain conjoined or connected across arbitrarily large space-time intervals, H. P. Stapp writes: "Everything we know about Nature is in accord with the idea that the fundamental process of Nature lies outside space-time (surveys the space-time continuum globally), but generates events that can be located in spacetime [*sic*]." "Are Superluminal Connections Necessary?" *Nuovo Cimento* 40B (1977): 202.

4. Stephen Toulmin, *The Philosophy of Science: An Introduction* (New York: Harper Torchbooks, 1960), p. 20. Compare Rom Harré, *The Philosophies of Science: An Introductory Survey* (London: Oxford University Press, 1972), p. 63: "In trying to understand the way images are formed by curved mirrors and lenses, and how what one sees in a plane mirror is related to its position, in trying to understand the way shadows appear, and in a host of other phenomena, it is found very convenient to suppose that something passes from the object through the glass of the lens or mirror, is reflected by the mirrored surface, or bent in its path by the lens, and finally reaches the eye. Whatever it is travels in straight lines. These lines are light rays. They are the basis of geometrical optics. Once they were thought to be the paths of corpuscles. Later they were supposed to be geometrical abstractions from moving wave fronts, and nowadays they retain something of each of these classical conceptions. But they are not seen in nature, they are drawn on paper."

5. Toulmin, *The Philosophy of Science*, p. 21.

6. Vasco Ronchi, *Optics: The Science of Vision*, trans. E. Rosen (Washington Square: New York University Press, 1957), p. 271. Emphasis in the original.

7. Geoffrey Cantor, *The Discourse of Light from the Middle Ages to the Enlightenment* (Berkeley and Los Angeles: University of California Press, 1985), pp. 96–97.

8. Bridgman, *The Logic of Modern Physics*, p. 153; see also Bridgman, *The Nature of Physical Theory* (New York: John Wiley and Sons, 1964), p. 126.

9. As Bridgman puts it: "There is no physical phenomenon whatever by which light may be detected apart from the phenomena of the source [emission apparatus] and the sink [absorption or detection apparatus]." *The Logic of Modern Physics*, p. 153.

10. In this regard, Bridgman states: "Physically it is the essence of light that it is *not* a thing that travels, and in choosing to treat it as a thing that does, I do not see how we can expect to avoid the most serious difficulties. . . . The properties of light remain incongruous and inconsistent when we try to think of them in terms of material things." *The Logic of Modern Physics*, p. 164; emphasis in the original.

11. Wheeler, *A Journey into Gravity and Spacetime*, p. 43.

12. Edwin F. Taylor and John A. Wheeler, *Spacetime Physics* (New York: W. H. Freeman, 1966), pp. 31–32.

13. James J. Gibson, *The Senses Considered as Perceptual Systems* (Boston: Houghton Mifflin Co., 1966), p. 2.

14. Ibid., p. 17.

15. Ibid., p. 226.

16. Ibid., p. 198.

17. Ibid., p. 3.

18. Ibid.

19. Ibid., p. 198.

20. Gibson, *The Ecological Approach to Visual Perception* (Boston: Houghton Mifflin, 1979), p. 3.

21. Quoted in Edward S. Reed, *James J. Gibson and the Psychology of Perception* (New Haven, Conn.: Yale University Press, 1988), p. 257. Emphasis in the original.

22. James J. Gibson, "A Note on the Relation Between Perceptual and Conceptual Knowledge." Unpublished manuscript (July 1974) at http://lor.trincoll.edu/~psyc/perils/folder6/perceptcon.html (accessed 5 April 2004). Emphasis in the original. See also James J. Gibson, "Wave-Train Information and Wave-Front Information in Sound and Light, With a Note on Ecological Optics." Unpublished manuscript (September 1968) at http://www.huwi.org/gibson/wavetrain.php (accessed 5 April 2004).

23. Claire F. Michaels and Claudia Carello, *Direct Perception* (Englewood Cliffs, N.J.: Prentice-Hall, 1981), pp. 179–80.

24. Ibid., p. 181.

25. Since nothing can move faster than light, writes John Schumacher, "nothing, not even light itself, can bring us news of its upcoming arrival." *Human Posture*, pp. 113–14.

26. Aristotle proposed that knowing what metaphors control one's thinking is the first step toward wider understanding: "It is a great thing, indeed, to make proper use of these poetical forms, as also of compounds and strange words. But the greatest thing by far is to be a master of metaphor. It is the one thing that cannot be learnt from others; and it is also a sign of genius, since a good metaphor implies an intuitive perception of the similarities in dissimilars." *De Poetica*, in *The Basic Works of Aristotle*,

ed. Richard Mckeon (New York: Modern Library, 2001), p. 1479. For the ubiquity of metaphors in human thought, see Paul Ricoeur, *The Rule of Metaphor: Multidisciplinary Studies of the Creation of Meaning in Language* (Toronto: University of Toronto Press, 1981).

27. Thomas Carlyle, *Past and Present* (N.p.: Chapman and Hall, Limited, 1897), book 2, chapter 17, p. 130.

28. See Bohr's remarks in Werner Heisenberg, *Physics and Beyond: Encounters and Conversations* (New York: Harper and Row, 1971), p. 41.

9. Pre-reflective Experience

1. Quoted in Des MacHale, *Wisdom* (London: Prion Books, 2002).

2. Albert Einstein, "Autobiographical Notes," trans. and ed. Paul Schilpp (La Salle, Ill.: Open Court, 1991), p. 5.

3. Johann Wolfgang von Goethe, *Theory of Colours*, trans. Charles Lock Eastlake (London: John Murray: 1840), p. 300.

4. Johann Wolfgang von Goethe, *Goethe: Die Schriften zur Naturwissenschaften*, part 1, vols. 4, 8 (Weimar: Hermann Böhlhaus Nachfolger, 1955), p. 18.

5. Ibid.

6. Emmanuel Levinas, *Discovering Existence with Husserl*, trans. Richard A. Cohen and Michael B. Smith (Evanston, Ill.: Northwestern University Press, 1998), p. 131.

7. Harida Chaudhuri, *Being, Evolution, and Immortality: An Outline of Integral Philosophy* (Wheaton, Ill.: Theosophical Publishing House, 1974), p. 195.

8. Plato, *The Republic*, p. 256. Translation modified: the pronoun "it" (referencing the sun) replaces "him" in Jowett's translation, which is mindful of Greek's gendered nouns.

9. René Descartes, *Discourse on Method*, part 4, in *Discourse on Method and Meditations on First Philosophy*, trans. Donald A. Cress, 4th ed. (Indianapolis: Hackett Publishing, 1998), p. 18.

10. Edmund Husserl, *The Essential Husserl: Basic Writings in Transcendental Phenomenology*, ed. Donn Welton (Bloomington: Indiana University Press, 1970), p. 364.

11. Husserl, *The Crisis of European Sciences and Transcendental Phenomenology*, p. 139.

12. Scott Clark, *Japan*, p. 7.

13. Robert Sokolowski, *Introduction to Phenomenology* (Cambridge: Cambridge University Press, 2000), p. 118.

10. Body, World, and Light

1. Maurice Merleau-Ponty, *Phenomenology of Perception*, trans. Colin Smith (London: Routledge, 1999), p. 67. Part of this chapter is taken from my "Merleau-Ponty's Visual Space and the Law of Large Numbers," *Studia Phænomenologica* 6 (2006): 391–406.

2. Ibid., p. 69.

3. Ibid., p. 68.

4. Ibid., pp. 68–69.

5. Harré, *The Philosophies of Science*, p. 63.

6. Merleau-Ponty, *Phenomenology of Perception*, p. 243. What he has in mind is that our body parts are not arrayed in space but enveloped, interrelated, and felt throughout; hence they constitute a very different kind of space from that idealized by science wherein things may be viewed dispassionately. See ibid., p. 98.

7. Merleau-Ponty, "The Eye and Mind," p. 187. Emphasis in the original.

8. Merleau-Ponty, *The Visible and the Invisible*, p. 273.

9. Merleau-Ponty, "The Eye and Mind," p. 187. Later Merleau-Ponty wrote: "he before whom the horizon opens is caught up, included within it. His body and the distances participate in one same corporeity or visibility in general, which reigns between them and it, and even beyond the horizon, beneath his skin, unto the depths of being." *The Visible and the Invisible*, p. 149.

10. Merleau-Ponty, *Phenomenology of Perception*, p. 216.

11. Ibid., p. 148.

12. Milič Čapek, *The Philosophical Impact of Contemporary Physics* (Princeton, N.J.: D. Van Nostrand, 1961), p. 184.

13. Merleau-Ponty, *The Visible and the Invisible*, p. 152.

14. Maurice Merleau-Ponty, *Husserl at the Limits of Phenomenology*, ed. and trans. Leonard Lawlor with Bettina Bergo (Evanston, Ill.: Northwestern University Press, 2002), p. x.

15. Merleau-Ponty, *The Visible and the Invisible*, p. 248. Emphasis in the original.

16. Wallace Stevens, "Six Significant Landscapes," *The Collected Poems of Wallace Stevens* (New York: Alfred A. Knopf, 1968), p. 74.

17. Merleau-Ponty, *The Visible and the Invisible*, p. 139.

18. Merleau-Ponty remarked that notwithstanding the immediacy of (pre-reflective) visual experience, space as interval or distance kicks in once that immediacy is noted: "The world is what I perceive, but as soon as we examine and express its absolute proximity, it also becomes, inexplicably, irremediable distance." *The Visible and the Invisible*, p. 8.

19. Merleau-Ponty, "The Eye and Mind," p. 187.

20. James, *Psychology*, p. 14. Emphasis in the original.

21. Merleau-Ponty, "The Eye and Mind," p. 166.

22. Ibid., p. 178.

23. Maurice Merleau-Ponty, *Themes from the Lectures at the Collège de France 1952–1960*, trans. J. O'Neill (Evanston, Ill.: Northwestern University Press, 1970), p. 86. Elsewhere Merleau-Ponty portrays Einstein as a classical thinker who did not realize that his revolutionary insights sprang from a shared world of pre-scientific perceptual experience. See Maurice Merleau-Ponty, *Signs*, trans. R. C. McCleary (Evanston, Ill.: Northwestern

University Press, 1964), pp. 192–97. Also, *The Visible and the Invisible*, pp. 14–27.

24. Merleau-Ponty, *Phenomenology of Perception*, p. 311.

25. Ibid., p. 310. Emphasis in the original. See also page 326: "My gaze 'knows' the significance of a certain patch of light in a certain context; it understands the logic of lighting."

26. Ibid., p. 310. Emphasis in the original.

27. Merleau-Ponty, *The Visible and the Invisible*, p. 151. Emphasis in the original.

28. Ibid., p. 135.

29. Ibid., p. 138.

30. Merleau-Ponty, "The Eye and Mind," p. 187.

31. Taylor and Wheeler, *Spacetime Physics*, p. 38.

32. Ibid.

33. Ibid.

34. Ibid.

35. Schumacher, *Human Posture*, p. 113. Emphasis in the original.

36. Ibid.

37. Ibid.

38. Ibid., 114.

39. Wheeler, *A Journey into Gravity and Spacetime*, p. 43.

40. Schumacher, *Human Posture*, p. 77.

41. After insisting that "We must rid ourselves of the illusion, encouraged by physics, that the perceived world is made up of color qualities," Merleau-Ponty states that "perception goes straight to the thing and bypasses the colour, just as it is able to fasten upon the expression of a gaze without noting the colour of the eyes" (Merleau-Ponty, *Phenomenology of Perception*, p. 305). He is proposing that colors (and, presumably, colorlessness) are not freestanding qualities but products of our involvement in the world.

42. See Galen A. Johnson, "The Problem of Origins: In the Timber Yard, Under the Sea," *Chiasmi International* 2 (2000): 249–57, for an interesting meditation on Merleau-Ponty's view of water, as intimated in *Eye and Mind*. While Johnson flirts with the possibility that Merleau-Ponty saw water as an escape from chiasmic indeterminacy (that perhaps explains its near-invisibility) he nevertheless concludes that water, like flesh, is ruptured in a self-generating way by its own non-self-identity. He states that "all that Merleau-Ponty wrote about vision and visibility was written about water" (p. 253).

43. Eliot, "The Dry Salvages," in *The Complete Poems and Plays*, pp. 130–37.

44. Zajonc, *Catching the Light*, p. 299.

45. Merleau-Ponty, *Phenomenology of Perception*, p. xvi.

11. Existential Light

1. Martin Heidegger, *Gesamtausgabe*, vol. 29/30 (Frankfurt am Main: Vittorio Klostermann, 1983), p. 519.

2. Michael Zimmerman, "Heidegger, Buddhism, and Deep Ecology," in *The Cambridge Companion to Heidegger*, ed. Charles B. Guignon (Cambridge: Cambridge University Press, 1993), pp. 244–245.

3. Hubert L. Dreyfus, *Being-in-the-World: A Commentary on Heidegger's* Being and Time (Cambridge, Mass.: MIT Press, 1991), pp. 245, 252.

4. Martin Heidegger, *Introduction to Metaphysics*, trans. Gregory Fried and Richard Polt (New Haven, Conn.: Yale University Press), pp. 107–108.

5. Blumenberg, "Light as a Metaphor for Truth," p. 31.

6. Heidegger, *Introduction to Metaphysics*, p. 16.

7. Ibid., p. 20.

8. Ibid., pp. 108–109.

9. James, *Psychology*, p. 14. Italics in the original.

10. Frederick Olafson, *Heidegger and the Philosophy of Mind* (New Haven, Conn.: Yale University Press, 1987), p. 226.

11. Theodore R. Schatzki, "Early Heidegger on Being, The Clearing, and Realism," *Heidegger Reexamined*, ed. Hubert Dreyfus and Mark Wrathall (New York: Routledge, 2002), vol. 2, pp. 177–194. The Heidegger citation is from *Being and Time*, trans. John Macquarrie and Edward Robinson (Oxford: Basil Blackwell, 1962), p. 27.

12. Schatzki, "Early Heidegger," p. 179.

13. Ibid., p. 185.

14. Dreyfus, *Being-in-the-World*, p. 245.

15. Zajonc, *Catching the Light*, p. 299.

16. Galilei, *Dialogue on the Great World Systems*, pp. 378–79.

17. Bohm, *Wholeness and the Implicate Order*, p. 149. Emphasis in the original.

18. Henri Bortoft, "Counterfeit and Authentic Wholes: Finding a Means for Dwelling in Nature," in *Goethe's Way of Science: A Phenomenology of Nature*, ed. David Seamon and Arthur Zajonc (Albany: SUNY Press, 1998), p. 279.

19. Ibid.

20. John Locke, *An Essay Concerning Human Understanding*, in *Great Books of the Western World*, ed. R. M. Hutchins, vol. 35 (Chicago: William Benton, 1952), p. 87.

21. Aristotle, *De Anima* 429a13–24, trans. Lynne Ballew, in *Straight and Circular: A Study of Imagery in Greek Philosophy* (Assen, Neth.: Van Gorcum, 1979), p. 128.

22. Ballew, *Straight and Circular*, p. 128.

23. Mikkel Borch-Jacobsen, *Lacan, The Absolute Master*, trans. Douglas Brick (Stanford, Calif.: Stanford University Press, 1991), p. 53.

24. E. A. Wallis Budge, "The Legend of Ra and Isis," in *The Gods of the Egyptians, Or Studies in Egyptian Mythology* (New York: Dover, 1969), vol. 1, p. 383.

25. John A. Wilson, "Egypt: The Nature of the Universe," in Henri Frankfort, et al., *The Intellectual Adventure of Ancient Man* (Chicago: University of Chicago Press, 1946), pp. 60–61.

26. "The Memphite Theology of Creation," translation and commentary by John A. Wilson, in *The Ancient Near East: An Anthology of Texts and Pictures*, ed. James B. Pritchard (Princeton, N.J.: Princeton University Press, 1958), vol. 1, p. 1.

27. Speech and song, argues Walter Otto, once were considered modes of revelation about the world, not merely means of self-expression. *Die Musen und der goettliche Ursprung des Singen und Sagens* (Düsseldorf: E. Diedrich, 1955), pp. 71–88.

28. Victor Zuckerkandl, *Man the Musician* (Princeton, N.J.: Princeton University Press, 1973), p. 12.

29. George Steiner, *Real Presences* (Chicago: University of Chicago Press, 1991), pp. 93–94, 105.

30. Thomas Sheehan, "*Kehre* and *Ereignis*: A Prolegomenon to *Introduction to Metaphysics*," in *A Companion to Heidegger's* Introduction to Metaphysics, ed. Gregory Fried and Richard Polt (New Haven, Conn.: Yale University Press), pp. 3–16.

31. Michael Welker, "Creation: Big Bang or the Work of Seven Days?" *Theology Today* 52, no. 2 (July 1995): 186.

32. R. T. Rundle Clark, *Myth and Symbol in Ancient Egypt* (London: Thames and Hudson, 1991), p. 227.

33. Ibid., p. 223.

34. Ibid., p. 90.

35. Acts 17:28, King James Version.

36. Joel R. Primack and Nancy Ellen Abrams, "'In a Beginning . . .' Quantum Cosmology and Kabbalah," *Tikkun* 10, no. 1 (Jan./Feb. 1995): 66–73.

37. Ibid., p. 73.

38. Ibid., p. 70.

39. Ibid., p. 72.

40. Ibid., p. 70

41. Blumenberg, "Light as a Metaphor for Truth," p. 31.

42. John Calvin, *Institutes of the Christian Religion*, ed. John T. McNeill and trans. Ford Lewis Battles (Philadelphia: Westminster Press, 1960), book 2, chapter 13, pt. 2.

43. Sheehan, "*Kehre* and *Ereignis*," p. 11.

44. Lord Kelvin, *Baltimore Lectures on Molecular Dynamics and the Wave Theory of Light* (London: Cambridge University Press Warehouse, 1904), p. 487.

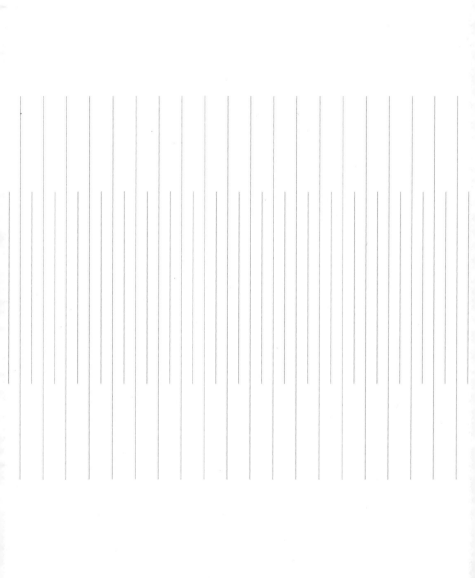

BIBLIOGRAPHY

Alighieri, Dante. *Paradiso.* In *The Divine Comedy of Dante Alighieri, A Verse Translation: Paradiso.* Trans. Allen Mandelbaum. Berkeley and Los Angeles: University of California Press, 1982.

Aristotle. *De Anima* 429a13–24. Quoted and trans. Lynne Ballew, *Straight and Circular: A Study of Imagery in Greek Philosophy.* Assen, Neth.: Van Gorcum, 1979.

———. *De Poetica.* In *The Basic Works of Aristotle.* Ed. Richard Mckeon. New York: Modern Library, 2001.

———. *Metaphysics.* Trans. Hippocrates G. Apostle. Bloomington: Indiana University Press, 1966.

Baierlein, Ralph. *Newton to Einstein: The Trail of Light.* Cambridge: Cambridge University Press, 2001.

Ball, Philip. *The Self-Made Tapestry: Pattern Formation in Nature.* Oxford: Oxford University Press, 1999.

Barrow, John D. *Pi in the Sky: Counting, Thinking, and Being.* Oxford: Clarendon Press, 1992.

Blackmore, Richard. *Creation. A Philosophical Poem* (London, 1712), bk ii., pp. 386–89. Quoted in G. N. Cantor, *Optics after Newton: Theories of Light in Britain and Ireland, 1704–1840,* epigraph preceding title page. Manchester, UK: Manchester University Press, 1983.

Blumenberg, Hans. 1993. "Light as a Metaphor for Truth." In *Modernity and the Hegemony of Vision,* ed. David Michael Levin, 30–86. Berkeley and Los Angeles: University of California Press.

Bohm, David. *Quantum Theory.* Englewood Cliffs, N.J.: Prentice Hall, 1951.

———. *Wholeness and the Implicate Order.* London: Ark Paperbacks, 1983.

Bondi, Hermann. *Relativity and Common Sense: A New Approach to Einstein.* Garden City, N.Y.: Anchor Books, 1964.

Boorstin, Daniel J. *The Creators: A History of Heroes of the Imagination.* New York: Vintage, 1992.

Borch-Jacobsen, Mikkel. *Lacan, The Absolute Master.* Trans. Douglas Brick. Stanford, Calif.: Stanford University Press, 1991.

Bortoft, Henri. "Counterfeit and Authentic Wholes: Finding a Means for Dwelling in Nature." In *Goethe's Way of Science: A Phenomenology of Nature,* ed. David Seamon and Arthur Zajonc. Albany: SUNY Press, 1998.

Bridgman, P. W. *The Logic of Modern Physics.* New York: Macmillan, 1955.

———. *The Nature of Physical Theory.* New York: John Wiley and Sons, 1964.

Budge, E. A. Wallis. "The Legend of Ra and Isis." In *The Gods of the Egyptians, Or Studies in Egyptian Mythology.* Vol. 1. New York: Dover, 1969.

Calvin, John. *Institutes of the Christian Religion.* Ed. John T. McNeill and trans. Ford Lewis Battles. Philadelphia: Westminster Press, 1960.

Cantor, G. [Geoffrey] N. *The Discourse of Light from the Middle Ages to the Enlightenment.* Berkeley and Los Angeles: University of California Press, 1985.

———. *Optics after Newton: Theories of Light in Britain and Ireland, 1704–1840.* Manchester, UK: Manchester University Press, 1983.

Čapek, Milič. *The Philosophical Impact of Contemporary Physics.* New York: D. Van Nostrand, 1961.

Carlyle, Thomas. *Past and Present.* N.p: Chapman and Hall, 1897.

Chaudhuri, Harida. *Being, Evolution, and Immortality: An Outline of Integral Philosophy.* Wheaton, Ill.: Theosophical Publishing House, 1974.

Clark, R. T. Rundle. *Myth and Symbol in Ancient Egypt.* London: Thames and Hudson, 1991.

Clark, Scott. *Japan: A View from the Bath.* Honolulu: University of Hawaii Press, 1994.

Derrida, Jacques. "Faith and Religion: The Two Sources of Religion at the Limits of Reason Alone." In *Religion,* ed. Jacques Derrida and Gianni Vattimo, trans. Samuel Weber. Stanford, Calif.: Stanford University Press, 1998.

———. *Writing and Difference.* Trans. Alan Bass. Chicago: University of Chicago Press, 1978.

Descartes, René. *Discourse on Method and Meditations on First Philosophy.* Trans. Donald A. Cress. 4th ed. Indianapolis: Hackett Publishing, 1998.

Dreyfus, Hubert L. *Being-in-the-World: A Commentary on Heidegger's Being and Time, Division I.* Cambridge, Mass.: MIT Press, 1991.

Dyson, Freeman. *Disturbing the Universe.* New York: Harper and Row, 1979.

The Egyptian Book of the Dead: The Papyrus of Ani. Trans. E. A. Budge. New York: Dover, 1967.

Einstein, Albert. *Autobiographical Notes.* Trans. and ed. Paul Schilpp. La Salle, Ill.: Open Court, 1991.

———. "On the Electrodynamics of Moving Bodies." In Arthur Miller, *Albert Einstein's Special Theory of Relativity*. Reading, Mass.: Addison-Wesley, 1981.

Eliot, T. S. *The Complete Poems and Plays: 1909–1950*. New York: Harcourt, Brace, and World, 1952.

Fagg, Lawrence W. *Electromagnetism and the Sacred: At the Frontier of Spirit and Matter*. New York: Continuum, 1999.

Fraser, J. T. *The Genesis and Evolution of Time: A Critique of Interpretation in Physics*. Amherst: University of Massachusetts Press, 1982.

———. *Time: The Familiar Stranger*. Amherst: University of Massachusetts Press, 1987.

Fritzsch, Harald. *An Equation That Changed the World: Newton, Einstein, and the Theory of Relativity*. Trans. Karin Heusch. Chicago: University of Chicago Press, 1994.

Galilei, Galileo. *Dialogue on the Great World Systems*. In the Salusbury Translation, revised, annotated, and with an introduction by Giorgio de Santillana. Chicago: University of Chicago Press, 1953.

Genz, Henning. *Nothingness: The Science of Empty Space*. Trans. Karin Heusch. Reading, Mass.: Helix Books, 1994.

Gibson, James J. *The Ecological Approach to Visual Perception*. Boston: Houghton Mifflin, 1979.

———. "A Note on the Relation Between Perceptual and Conceptual Knowledge." July 1974. Unpublished manuscript at http://lor.trincoll.edu/~psyc/perils/folder6/perceptcon.html (accessed 5 April 2004).

———. *The Senses Considered as Perceptual Systems*. Boston: Houghton Mifflin, 1966.

———. "Wave-Train Information and Wave-Front Information in Sound and Light, With a Note on Ecological Optics." September 1968. Unpublished manuscript at http://www.huwi.org/gibson/wavetrain.php (accessed 5 April 2004).

Goethe, Johann Wolfgang von. *Faust, Parts One and Two*. Trans. George Madison Priest. Chicago: William Benton, 1952.

———. *Goethe: Die Schriften zur Naturwissenschaften*, Weimar: Hermann Böhlhaus Nachfolger, 1955.

———. *Theory of Colours*. Trans. Charles Lock Eastlake. London: John Murray, 1840.

Grandy, David. "Light as a Solution to Puzzles about Light." In *Journal for General Philosophy of Science* 33 (2002): 369–79.

———. "Merleau-Ponty's Visual Space and the Law of Large Numbers." In *Studia Phænomenologica* 6 (2006): 391–406.

Greenstein, George, and Arthur Zajonc. *The Quantum Challenge: Modern Research on the Foundations of Quantum Mechanics*. Boston: Jones and Bartlett, 1997.

Grosseteste, Robert. *On the Six Days of Creation: A Translation of the Hexaëmeron*. Trans. C. F. J. Martin. Oxford: Oxford University Press, 1996.

Haisch, Bernard. "Brilliant Disguise: Light, Matter and the Zero-Point Field." *Science and Spirit* 10, no. 3 (1999): 30–31.

Harré, Rom. *The Philosophies of Science: An Introductory Survey.* London: Oxford University Press, 1972.

Hawking, Stephen W. "The Beginning of Time." http://www.hawking.org .uk/lectures/bot.html (accessed 26 June 2005).

Heiddeger, Martin. *Being and Time.* Trans. John Macquarrie and Edward Robinson. Oxford: Basil Blackwell, 1962.

———. *Gesamtausgabe,* vol. 29/30. Frankfurt am Main: Vittorio Klostermann, 1983.

———. *Introduction to Metaphysics.* Trans. Gregory Fried and Richard Polt. New Haven, Conn.: Yale University Press, 2000.

Heisenberg, Werner. *The Physicist's Conception of Nature.* New York: Harcourt, Brace, 1958.

———. *Physics and Beyond: Encounters and Conversations.* New York: Harper and Row, 1971.

———. *Physics and Philosophy.* New York: Prometheus, 1999.

Hoffman, Banesh. *Relativity and Its Roots.* New York: Scientific American Books, 1983.

Hopkins, Gerard Manley. *The Major Poems.* Ed. Walford Davies. London: J. M. Dent and Sons, 1979.

Husserl, Edmund. *The Crisis of European Sciences and Transcendental Phenomenology: An Introduction to Phenomenology.* Trans. David Carr. Evanston, Ill.: Northwestern University Press, 1970.

———. *The Essential Husserl: Basic Writings in Transcendental Phenomenology.* Ed. Donn Welton. Bloomington: Indiana University Press, 1970.

Jacobsen, Thorkild. "Mesopotamia: The Cosmos as a State." In Henri Frankfort, et al., *The Intellectual Adventure of Ancient Man,* 125–184. Chicago: University of Chicago Press, 1946.

Jaki, Stanley L. *Is There a Universe?* Liverpool, UK: Liverpool University Press, 1993.

James, William. *Psychology.* New York: Henry Holt, 1910.

———. *Some Problems of Philosophy: A Beginning of an Introduction to Philosophy.* Lincoln: University of Nebraska Press, 1996.

———. *Varieties of Religious Experience.* New York: Vintage Books, 1987.

Jammer, Max. *The Conceptual Development of Quantum Mechanics.* New York: McGraw-Hill, 1966.

Johnson, Galen A. "The Problem of Origins: In the Timber Yard, Under the Sea." *Chiasmi International* 2 (2000): 249–57.

Jonas, Hans. "The Nobility of Sight." In *The Phenomenon of Life: Toward a Philosophical Biology.* Chicago: University of Chicago Press, 1982.

Kelvin, Lord. *Baltimore Lectures on Molecular Dynamics and the Wave Theory of Light.* London: Cambridge University Press Warehouse, 1904.

Kirk, G. S., J. E. Raven, and M. Schofield. *The Presocratic Philosophers: A Critical History with a Selection of Texts,* 2nd ed. Cambridge: Cambridge University Press, 1983.

Kosso, Peter. *Appearance and Reality: An Introduction to the Philosophy of Physics.* New York: Oxford University Press, 1998.

Koyré, Alexandre. *From the Closed World to the Infinite Universe.* Baltimore, Md.: Johns Hopkins Press, 1968.

Levinas, Emmanuel. *Discovering Existence with Husserl.* Trans. Richard A. Cohen and Michael B. Smith. Evanston, Ill.: Northwestern University Press, 1988.

Locke, John. *An Essay Concerning Human Understanding.* In *Great Books of the Western World,* ed. R. M. Hutchins, vol. 35. Chicago: William Benton, 1952.

MacHale, Des. *Wisdom.* London: Prion Books, 2002.

MacKenzie, Iaian. *The "Obscurism" of Light: A Theological Study into the Nature of Light.* Norwich, UK: Canterbury Press, 1966.

Malin, Shimon. *Nature Loves to Hide: Quantum Physics and the Nature of Reality, a Western Perspective.* New York: Oxford University Press, 2001.

Merleau-Ponty, Maurice. "Eye and Mind." Trans. Carleton Dallery. In *The Primacy of Perception,* 159–90. Evanston, Ill.: Northwestern University Press, 1964.

———. *Husserl at the Limits of Phenomenology.* Ed. and trans. Leonard Lawlor with Bettina Bergo. Evanston, Ill.: Northwestern University Press, 2002.

———. *Phenomenology of Perception.* Trans. Colin Smith. London: Routledge, 1999.

———. *Signs.* Trans. R. C. McCleary. Evanston, Ill.: Northwestern University Press, 1964.

———. *Themes from the Lectures at the Collège de France 1952–1960.* Trans. J. O'Neill. Evanston, Ill.: Northwestern University Press, 1970.

———. *The Visible and the Invisible.* Trans. Alphonso Lingis. Evanston, Ill.: Northwestern University Press, 1968.

Mermin, N. David. *It's About Time: Understanding Einstein's Relativity.* Princeton, N.J.: Princeton University Press, 2005.

Michaels, Claire F., and Claudia Carello. *Direct Perception.* Englewood Cliffs, N.J.: Prentice-Hall, 1981.

Misner, Charles W., Kip S. Thorne, and John Archibald Wheeler. *Gravitation.* San Francisco: W. H. Freeman, 1973.

Nadeau, Robert, and Menas Kafatos. *The Non-local Universe: The New Physics and Matters of the Mind.* Oxford: Oxford University Press, 2001.

Natsoulas, Thomas. "Gibson's Environment, Husserl's Lebenswelt, the World of Physics, and the Rejection of Phenomenal Objects." *American Journal of Psychology* 107, no. 3 (1994): 327–58.

Newton, Isaac. *Principia Mathematica.* Trans. Andrew Motte (1729) as revised by Florian Cajori. Berkeley and Los Angeles: University of California Press, 1934.

Nibley, Hugh. *The Message of the Joseph Smith Papyrus: An Egyptian Endowment.* Salt Lake City, Utah: Deseret Book, 1975.

Olafson, Frederick. *Heidegger and the Philosophy of Mind.* New Haven, Conn.: Yale University Press, 1987.

Otto, Walter. *Die Musen und der goettliche Ursprung des Singen und Sagens.* Düsseldorf: E. Diedrich, 1955.

Palmer, Stephen E. *Vision Science: Photons to Phenomenology.* Cambridge, Mass.: MIT Press, 1999.

Perkowitz, Sydney. *Empire of Light: A History of Discovery in Science and Art.* New York: Henry Holt, 1996.

Plato. *The Republic.* Trans. Benjamin Jowett. New York: Vintage, 1991.

Powers, Jonathan. *Philosophy and the New Physics.* London: Methuen, 1982.

Primack, Joel R., and Nancy Ellen Abrams. "'In a Beginning . . .' Quantum Cosmology and Kabbalah." *Tikkun* 10, no. 1 (Jan./Feb. 1995): 66–73.

Reed, Edward S. *James J. Gibson and the Psychology of Perception.* New Haven, Conn.: Yale University Press, 1988.

Reichenbach, Hans. *From Copernicus to Einstein.* Trans. Ralph B. Winn. New York: Philosophical Library, 1942.

Ricoeur, Paul. *The Rule of Metaphor: Mult idisciplinary Studies of the Creation of Meaning in Language.* Toronto: University of Toronto Press, 1981.

Ronchi, Vasco. *Optics: The Science of Vision.* Trans. Edward Rosen. Washington Square: New York University Press, 1957.

Ruechardt, Eduard. *Light: Visible and Invisible.* Ann Arbor: University of Michigan Press, 1958.

Schatzki, Theodore R. "Early Heidegger on Being, The Clearing, and Realism." In *Heidegger Reexamined,* ed. Hubert Dreyfus and Mark Wrathall, 2:177–94. New York: Routledge, 2002.

Schumacher, John A. *Human Posture: The Nature of Inquiry.* Albany: SUNY Press, 1989.

Sheehan, Thomas. "*Kehre* and *Ereignis:* A Prolegomenon to *Introduction to Metaphysics.*" In *A Companion to Heidegger's* Introduction to Metaphysics, ed. Gregory Fried and Richard Polt, 3–16. New Haven, Conn.: Yale University Press, 2001.

Sheldrake, Rupert. *Seven Experiments That Could Change the World: A Do-It Yourself Guide to Revolutionary Science.* New York: Riverhead Books, 1995.

Smith, Barry. "Common Sense." In *The Cambridge Companion to Husserl,* ed. Barry Smith and David Woodruff Smith. Cambridge: Cambridge University Press, 1995.

Sokolowski, Robert. *Introduction to Phenomenology.* Cambridge: Cambridge University Press, 2000.

Stapp, H. P. "Are Superluminal Connections Necessary?" *Nuovo Cimento* 40B (1977): 191–205.

Steiner, George. *Real Presences.* Chicago: University of Chicago Press, 1991.

Stevens, Wallace. *The Collected Poems of Wallace Stevens.* New York: Alfred A. Knopf, 1968.

Taylor, Edwin F., and John A. Wheeler. *Spacetime Physics*. New York: W. H. Freeman, 1966.

Torrance, Thomas F. "The Theology of Light." In *Christian Theology and Scientific Culture*, comprising the Theological Lectures at the Queen's University, Belfast for 1980. Eugene, Ore.: Wipf and Stock, 1998.

Toulmin, Stephen. *The Philosophy of Science: An Introduction*. New York: Harper Torchbooks, 1960.

Welker, Michael. "Creation: Big Bang or the Work of Seven Days?" *Theology Today* 52, no. 2 (1995): 173–87.

Westfall, Richard S. *The Construction of Modern Science: Mechanisms and Mechanics*. New York: John Wiley and Sons, 1971.

———. *The Life of Isaac Newton*. Cambridge: Cambridge University Press, 1993.

Wheeler, John Archibald. *A Journey into Gravity and Spacetime*. New York: Scientific American Library, 1990.

———. "Law without Law." In *Quantum Theory and Measurement*, ed. John Archibald Wheeler and Wojciech Hubert Zurek, pp. 182–213. Princeton, N.J.: Princeton University Press, 1983.

Whitehead, Alfred North. *The Concept of Nature*. Mineola, N.Y.: Dover, 2004.

———. *Process and Reality*. Corrected ed. by David Ray Griffin and Donald W. Sherburne. New York: Free Press, 1985.

———. *Science and the Modern World*. New York: Free Press, 1967.

Wilson, John A. "Egypt: The Nature of the Universe." In Henri Frankfort, et al., *The Intellectual Adventure of Ancient Man*, 31–61. Chicago: University of Chicago Press, 1946.

———. "The Memphite Theology of Creation." Translation and commentary by John A. Wilson. In *The Ancient Near East: An Anthology of Texts and Pictures*, vol. 1, ed. James B. Pritchard, 1–2. Princeton, N.J.: Princeton University Press, 1958.

Wordsworth, William. *Wordsworth: Poetical Works*. Ed. Thomas Hutchinson. New ed., rev. Ernest De Selincourt. Oxford: Oxford University Press, 1978.

Zajonc, Arthur. *Catching the Light: The Entwined History of Light and Mind*. New York: Oxford University Press, 1993.

Zimmerman, Michael. "Heidegger, Buddhism, and Deep Ecology." In *The Cambridge Companion to Heidegger*, ed. Charles B. Guignon, pp. 240–69. Cambridge: Cambridge University Press, 1993.

Zuckerkandl, Victor. *Man the Musician*. Princeton, N.J.: Princeton University Press, 1973.

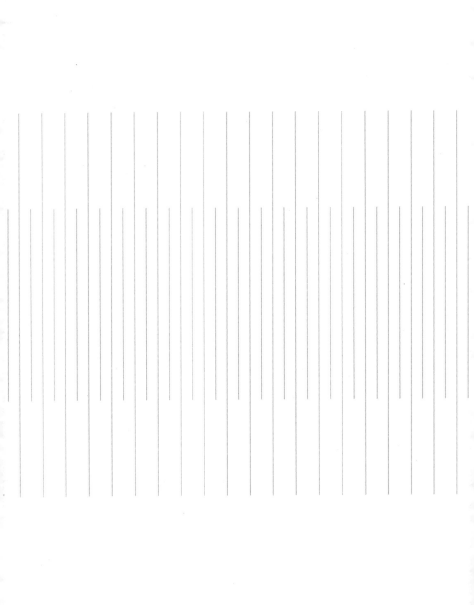

INDEX

Abrams, Nancy, 161–162, 180
air, 14, 18, 64, 142–143, 161
apartness, 41, 52–53, 72
Aristotle, 13–15, 94, 155, 175, 179
atom(s), 44, 53–54, 83, 104, 160
Augustine, 68
Australian Aborigine Creation
 Story, 12

Bailerlein, Ralph, 87
Ballew, Lynne, 179
Barrow, John D., 97
beam splitter, 74–75
being-in-the-world, 148–149, 155,
 164
big bang, 64–65, 102, 157, 162,
 167–168, 180; singularity of the,
 3, 9, 102, 165, 167–168
black body radiation, 164
Blackmore, Richard, 49
Blumenberg, Hans, 3, 163, 167,
 170, 179–180
Bohm, David, 72–76, 78, 153,
 169–170, 172, 176, 179; Bohm's
 explicate order, 72–73, 76;
 Bohm's implicate order, 76
Bohr, Niels, 112
Bondi, Hermann, 65, 171
Boorstin, Daniel, 171

Borch-Jacobsen, Mikkel, 179
Bortoft, Henri, 153–154, 179
Boyle's law, 134
brain, 72, 76, 102–103, 105–107,
 125, 135; brain-world continuity, 103; brain-world dissimilarity, 103
Bridgman, P. W., 101, 174–175
Budge, E. A. Wallis, 179

Calvin, John, 163, 180
Cantor, Geoffrey, 101, 174
Čapek, Milič, 44, 133, 170, 177
Carello, Claudia, 111, 175
Carlyle, Thomas, 124, 176
Chaudhuri, Harida, 176
Clark, Rundle, 157–159, 180
Clark, Scott, 172, 176
clock(s), 2, 27, 34, 38–39
color(s)/colour(s), 49, 108, 118, 122,
 138, 143, 173, 176, 178
consciousness, 83, 123–125, 144,
 156; divine, 156
constancy, 1–3, 14, 20, 22, 27, 30,
 37, 42, 46, 49–51, 65–66, 71,
 121–122, 134, 138, 150–151, 167;
 horizontal, 43–44; measured, 6,
 22, 26, 42; observed, 42
Copenhagen Interpretation, 45–46

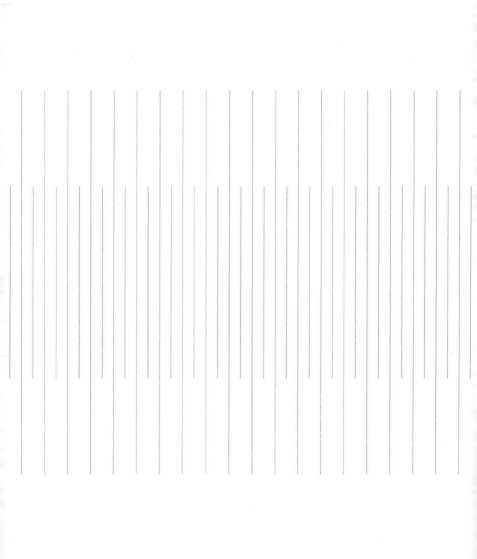

DAVID A. GRANDY is Professor of Philosophy at Brigham Young University and author (with Dan Burton) of *Magic, Mystery, and Science* (Indiana University Press, 2004) and *Leo Szilard: Science as a Mode of Being.*